Agricultural and Food Waste

Agricultural and Food Waste: Analysis, Characterization, and Extraction of Bioactive Compounds and Their Possible Utilization

Editors

Montserrat Dueñas Patón
Ignacio García-Estévez

MDPI • Basel • Beijing • Wuhan • Barcelona • Belgrade • Manchester • Tokyo • Cluj • Tianjin

Editors
Montserrat Dueñas Patón
Universidad de Salamanca
Spain

Ignacio García-Estévez
University of Salamanca Salamanca
Spain

Editorial Office
MDPI
St. Alban-Anlage 66
4052 Basel, Switzerland

This is a reprint of articles from the Special Issue published online in the open access journal *Foods* (ISSN 2304-8158) (available at: https://www.mdpi.com/journal/foods/special_issues/Agricultural_Food_Waste_Analysis_Characterization_Extraction_Bioactive_Compounds_Utilization).

For citation purposes, cite each article independently as indicated on the article page online and as indicated below:

LastName, A.A.; LastName, B.B.; LastName, C.C. Article Title. *Journal Name* **Year**, *Article Number*, Page Range.

ISBN 978-3-03943-346-9 (Hbk)
ISBN 978-3-03943-347-6 (PDF)

Contents

About the Editors

Montserrat Dueñas Patón holds a PhD in Chemistry from the University Autónoma from Madrid; she is a Professor at the University of Salamanca (Spain). She has developed part of her research in renowned institutions, such as Instituto de Fermentaciones Industriales (CSIC), Estaçao Vitivinicola Nacional (INIA-Portugal) and INRA Montpellier (France). She has authored over 100 scientific publications (research papers, reviews and book chapters).Her research work focuses on several main lines: a) characterization of phenolic compounds in foods; b) the application of technological processes in order to improve the quality of foods; c) bioavailability and biological activity of dietary flavonoids; and d) the relation of phenolic compounds with the organoleptic properties of wines (color and astringency).

Ignacio García-Estévez received in 2014 his PhD degree from the University of Salamanca. He is a postdoctoral researcher at the department of Food Science and Nutrition of the same university. He has carried out part of his research at the Institut Européen de Chimie et Biologie (University of Bordeaux, France) and Faculty of Chemistry of University of Porto (Portugal). He has authored more than 50 scientific publications (peer-reviewed papers, book chapters and reviews) related to his research work, which is focused on the following main lines: i) the relation between phenolic composition and wine organoleptic properties; ii) the study of phenolic composition in vegetable matrices and foods, and its use as chemotaxonomic markers; and iii) the evaluation of technological strategies for the modulation of wine organoleptic properties (color and astringency).

foods MDPI

Editorial

Agricultural and Food Waste: Analysis, Characterization and Extraction of Bioactive Compounds and Their Possible Utilization

Montserrat Dueñas * and Ignacio García-Estévez

Grupo de Investigación en Polifenoles, Departamento de Química Analítica, Nutrición y Bromatología, Facultad de Farmacia, Universidad de Salamanca, E37007 Salamanca, Spain; igarest@usal.es

* Correspondence: mduenas@usal.es; Tel.: +34-923-294-537

Received: 20 May 2020; Accepted: 15 June 2020; Published: 21 June 2020

Abstract: The characterization and reutilization of agricultural and food waste is an important strategy to ensure the sustainable development of the agricultural and food industries. As a result, the environmental impact of these industries can be reduced, thus contributing to the fight against environmental problems, mainly to those related to a potential mitigation of climatic change. This Special Issue includes five papers that reported important findings from research activities related to the reutilization of by-products from food processing industries, which help to increase the knowledge in this field

Keywords: bioactive compounds; food waste; functional foods; by-products; characterization and extraction; phytochemicals; climatic change; phenolic compounds

---

The food processing industries produce millions of tons of losses and waste during processing, which is becoming a grave economic, environmental and nutritional problem. Fruit, vegetable and food industrial solid waste includes several products that are released in food production during cleaning, processing, cooking, and/or packaging. These wastes can be an important source of bioactive compounds since, in their composition, important levels of phenolic compounds, dietary fibers, polysaccharides, vitamins, carotenoids, pigments and oils, among others, can be found. These compounds, in turn, are closely associated with beneficial effects in human health. Thus, these by-products can be exploited again in the food industry to develop functional ingredients and/or new foods or natural additives or, in other industries, such as the pharmaceutical, agricultural or chemical industries, to obtain cosmetic products, fertilizers or animal feed, among other things. Therefore, the characterization and reutilization of these by-products is important to ensure the sustainable development of the food industry and reduce its environmental impact, which would contribute to the fight against environmental problems, mainly to a potential mitigation of climatic change.

Among the submitted works, five papers have been selected to be included in this Special Issue. The study performed by Mislata and co-workers assesses the levels and identities of the aromatic and bioactive compounds, which, in turn, can be related to beneficial health effects, that can be found in the by-products of cork stopper production [1]. Thus, these authors reported the potential added value of the waste of that industry. To be precise, three different cork granulates that can be used to produce cork stoppers, which differed in their size, along with the corresponding cork stoppers, were analyzed in order to determine the phenolic and aromatic compositions, as well as the antioxidant activity of the corresponding extracts. The results indicated that several aromatic compounds with industrial interest can be extracted mainly from the cork by-products, among them, vanillins and volatile phenols such as 4-vinylguaiacol. The cork by-products also showed important concentrations of phenolic compounds, gallic and protocatechuic acids being the most important phenolic compounds extracted. These phenolic compounds are related by these authors to the high antioxidant activity determined in the extracts obtained from cork by-products, although aromatic compounds such as vanillin can also be responsible for that activity. Altogether, these authors reported the high potential of cork by-products and cork stoppers, but mainly the former, to be reused as flavoring agents and antioxidants in the food industry.

The study carried out by Jiang et al. is also included in this Special Issue. In this paper, an innovative methodology for the extraction and purification of ovalbumin from salted egg whites, obtained as by-products of the industrial obtention of salted egg yolks, is outlined [2]. This represents an important step for the reutilization of these by-products that are usually discarded as waste because of the difficulty in treatment and that can cause important environmental problems due to their salt content. The new methodology comprises an aqueous two-phase flotation that allows the separation of ovalbumin from the food by-products. This process, described as simple, inexpensive and efficient, was developed and optimized by using a response surface method experiment, and it allowed the authors to obtain a purified ovalbumin extract, which was characterized by means of different techniques such as SDS-PAGE, RP-HPLC, nano-LC-ESI-MS/MS, UV, fluorescence and FT-IR spectroscopy. This characterization reveals the high purity of the ovalbumin obtained, whose structure and properties in oil binding capacity, viscosity, emulsibility and foam capacity did not show differences with the corresponding standard. Thus, this new methodology can be considered as a sustainable and effective way for the utilization of salted egg whites to reduce environmental pollution.

This Special Issue also includes the study performed by F. Correddu et al. [3], in which the proximate composition and phenolic compounds content, as well as the antioxidant activity of *Myrtus communis* berries obtained as waste from myrtle-liqueur production, have been studied. These analyses were also carried out in the different parts of berries: seeds and pericarps. These berries are typically used to elaborate a sweet myrtle-liqueur by their hydroalcoholic infusion for at least two weeks. The results obtained by these authors demonstrated that exhausted myrtle berries presented a high concentration of carbohydrates, proteins and lipids, showing that the seeds have higher levels of hemicellulose, cellulose and lipids than pericarps and whole berries. The lipid fraction showed a high concentration of polyunsaturated fatty acids, linoleic acid being the most important fatty acid in seeds and whole exhausted myrtle berries. With regard to the phenolic profile, hydroxybenzoic and hydroxycinnamic acids and flavonols such as quercetin, isorhamnetin and kaempferol were identified in these berries. Ellagic acid was the one presented in the highest levels, followed by gallic acid. Among the flavonols, quercetin aglycon and quercetin 3-*O*-rutinoside were the most abundant flavonoids. This study showed that the seeds presented the highest total phenolic compounds concentration and, as a consequence, the highest antioxidant activity. Hence, this study suggests the importance of this by-product with multiple industrial applications, as a food ingredient or in animal feed formulations.

The study carried out by Vella et al. [4] is focused on total phenolic compounds, ortho-diphenol, flavonoid and tannin contents and antioxidant activity of peels and seeds from Cantaloupe melon. The Cantaloupe melon is the most cultivated variety in Italy and it is characterized by its high content of vitamins A and C and minerals such as potassium and magnesium [5]. The melon processing industries produce great quantities of waste during processing, mainly peel and seeds. These by-products could be exploited in order to minimize economic and environmental problems. The results reported suggest that peel and seeds of Cantaloupe melon are promising sources of natural phytochemicals, which could be related to their high polyphenol and tannin concentrations, mainly in the case of peels. These investigations would be useful for the food industry to develop new nutraceuticals, as well as to reduce waste volumes, thus contributing to the sustainable management of waste that involves environmental and economic costs.

S. Silbir and Y. Goksungur [6] have carried out studies on the re-utilization of brewery waste hydrolysate with the aim of obtaining natural red pigments by *Monascus purpureus* CMU001, used in the fermentation of brewer's spent grain. This study was mainly focused on the chemical, structural and elemental characterization of this by-product using different techniques such as Fourier-transform infrared spectroscopy (FT-IR) and X-Ray Photoelectron Spectroscopy (XPS). The obtained results show that the brewer's spent grain was constituted mainly by lignin, hemicellulose, cellulose and protein, whose presence was confirmed by infrared spectrum (FT-IR). The XPS analysis indicated the presence of carbon, oxygen, nitrogen and phosphorus. These authors also reported the optimization of the fermentation process to obtain the maximum natural red pigment concentration by *Monasuc purpureus*, reaching the highest biomass concentration at 7 days of fermentation, after which it declined. Therefore, this study suggests an important alternative for using this waste, produced in great quantities in breweries, and thus, to obtain natural pigments, which present multiple beneficial properties for health.

We are pleased to present this Special Issue, which includes five papers that highlight the most important of the research activities in the field of the reutilization of by-products from food processing industries, which could contribute to the mitigation of environmental problems, mainly in terms of their climatic change potential. We are very grateful to the authors who have shared their scientific knowledge and experience through their contribution to this Special Issue. We sincerely hope that the readers will find this Special Issue interesting and informative.

Funding: FEDER Interreg España-Portugal Programme: project ref 0377_IBERPHENOL_6_E; Strategic Research Programs for Units of Excellence from Junta de Castilla y León: ref CLU-2O18-04; Spanish MICIU (Ministry of Science, Innovation and Universities: AGL2017-84793-C2-1-R; the Spanish MICIU for the Juan de la Cierva-Incorporación postdoctoral contract: Grant IJCI-2017-31499.

Acknowledgments: We thank FEDER Interreg España-Portugal Programme (project ref 0377_IBERPHENOL_6_E), Strategic Research Programs for Units of Excellence from Junta de Castilla y León (ref CLU 2O18 04) and Spanish MICIU (Ministry of Science, Innovation and Universities, Project Reference AGL2017-84793-C2-1-R cofunded by FEDER). I.G.-E. thanks the Spanish MICIU for the Juan de la Cierva-Incorporación postdoctoral contract (Grant IJCI-2017-31499).

Conflicts of Interest: The authors declare no conflict of interest.

References

1. Mislata, A.M.; Puxeu, M.; Ferrer-Gallego, R. Aromatic potential and bioactivity of cork stoppers and cork by-products. *Foods* **2019**, *9*, 133. [CrossRef] [PubMed]
2. Jiang, B.; Na, J.; Wang, L.; Li, D.; Liu, C.; Feng, Z. Reutilization of food waste: One-step extration, purification and characterization of ovalbumin from salted egg white by aqueous two-phase flotation. *Foods* **2019**, *8*, 286. [CrossRef] [PubMed]
3. Correddu, F.; Maldini, M.; Addis, R.; Petretto, G.L.; Palomba, M.; Battacone, G.; Pulina, G.; Nudda, A.; Pintore, G. Myrtus communis liquor byproduct as a source of bioactive compounds. *Foods* **2019**, *8*, 237. [CrossRef] [PubMed]
4. Vella, F.M.; Cautela, D.; Laratta, B. Characterization of polypheolic compounds in Cantaloupe melón by-products. *Foods* **2019**, *8*, 196. [CrossRef] [PubMed]
5. Fundo, J.F.; Miller, F.A.; García, E.; Santos, J.R.; Silva, C.L.; Brandão, T.R. Physicochemical characteristics, bioactive compounds and antioxidant activity in juice, pulp, peel and seeds and Cantaloupe melon. *J. Food Meas. Charact.* **2018**, *12*, 292–300. [CrossRef]
6. Silbi, S.; Goksungur, Y. Natural red pigment production by Monascus Purpureus in submerged fermentations systems using a food industry waste: Brewer's spent grain. *Foods* **2019**, *8*, 161. [CrossRef] [PubMed]

foods

MDPI

Article

Aromatic Potential and Bioactivity of Cork Stoppers and Cork By-Products

Ana Maria Mislata [1,2], Miquel Puxeu [1] and Raul Ferrer-Gallego [1,*]

1 Centro Tecnológico del Vino—VITEC, Carretera de Porrera Km. 1, 43730 Falset, Spain; anamaria.mislata@vitec.wine (A.M.M.); miquel.puxeu@vitec.wine (M.P.)

2 Sensometria Instrumental (i-Sens), Department of Analytical Chemistry and Organic Chemistry, Universitat Rovira i Virgili, 43007 Tarragona, Spain

* Correspondence: raul.ferrer@vitec.wine

Received: 18 December 2019; Accepted: 23 January 2020; Published: 28 January 2020

Abstract: The characterization of natural waste sources is the first step on the reutilization process, circular economy, and global sustainability. In this work, the aromatic composition and bioactive compounds related to beneficial health effects from cork stoppers and cork by-products were assessed in order to add value to these wastes. Twenty-three aromatic compounds with industrial interest were quantified by gas chromatography coupled mass spectrometry GC–MS in both samples. Vanillins and volatile phenols were the most abundant aromatic families. Other aromatic compounds, such as aldehydes, lactones, terpenols, and alcohols, were also determined. Furthermore, the phenolic composition and the antioxidant activity were also evaluated. Overall, extracts showed high aromatic and antioxidant potential to be further used in different industrial fields. The recovery of these valuable compounds from cork stoppers and cork by-products helps to reuse them in agricultural, cosmetic, pharmaceutical, or food industries.

Keywords: cork; volatile compounds; antioxidant activity; polyphenols; aroma; waste

1. Introduction

Nowadays, a key issue in any field of research is the fight to curb climate change. This environmental awareness is of great importance in every society in order to achieve a sustainable environment considering the current human actions. The re-use and material recycling is a priority for waste management in the European Union (Directive 2008/98/EC on Waste). The characterization of natural waste sources is the first step on the reutilization process and the global sustainability. Bioactive compounds related to beneficial health effects and aromatics from by-products can be explored by food, agricultural, cosmetic, or pharmaceutical industries. Fragments, granulates, and powder from cork represent a large waste stream from cork processing [1,2]. They have been commonly used as combustion fuel, although they have also been employed in agriculture. Composted residues from cork were used as plant growth media to suppress plant diseases [3], and hydrological properties of substrates based on industrial cork residue have also been reported [4]. Recently, their adsorption properties, such as fining agent in wines, were stated [5], and some works showed that cork wastes are cost-effective and green alternatives to the retention of contaminants from water [6–9]. However, to the best of our knowledge, few works were reported regarding their revalorization for the food industry. According to statistical reports of the International Organization of Vine and Wine (OIV), the annual world wine production is around 275 million hectoliters (292 MhL in 2018) [10], and 90% of bottle wines are stopped by corks. In cork stoppers production (about 300 thousand tons of cork are produced annually), the cork waste represents around 25% of the raw material. Different cork wastes can be found depending on their characteristics, density, moisture, granulometry, size, ash content, and tannin concentration [1,11,12].

It is well known that the phenolic composition of plants is related to antioxidative, anticarcinogenic, and antitumor biological activities [13,14]. Specially, polyphenols from *Quercus suber* L. have been associated with beneficial health effects linked to hydrolysable tannins and phenolic compounds with low molecular weight [15]. Recent research valorized cork powder and granulates from a phenolic point of view [12], however, little research focused on the recovery of valuable aromatic compounds. On the other hand, a great number of works are related to the off-flavors or cork-taint compounds in wine [16–19]. However, the characterization of the positive aromas of cork stoppers and cork by-products is scarce. In this work, the aromatic composition of different wine cork stoppers and granulates was determined, enhancing the added value of their re-use in other industrial fields. The tannin concentration and the antioxidant activity of the cork extracts were also compared and discussed.

2. Materials and Methods

2.1. Samples and Extraction

Different cork stoppers and their respective cork granulate (A, B, and C), from which the corks were made, were evaluated in this study. The granulate cork size ranged from 6.1 to 14.3 mm for A samples (high size), from 1.6 to 4.4 mm (medium size) for B samples, and from 0.85 to 1.9 mm (low size) for C samples (Figure 1 and Table S1). The cork stoppers dimensions were 30 × 50 mm for all samples. To determine the migration of the phenolic compounds from granulates and cork samples to model solution, a liquid–liquid extraction with ethyl acetate was made according to Azevedo et al. [20]. Briefly, 30 g of each sample (cork stoppers and granulates) was weighted in glass jars and macerated in 1 L of hydro alcoholic solution. The model solution contained 13% of ethanol, 5 g/L of tartaric acid, and the pH was adjusted to 3.7 with sodium hydroxide (1 *N*). Solutions were stored at controlled temperature (20–22 °C) until the analyses were performed. Different times of maceration were evaluated in this study (3, 5, and 15 days).

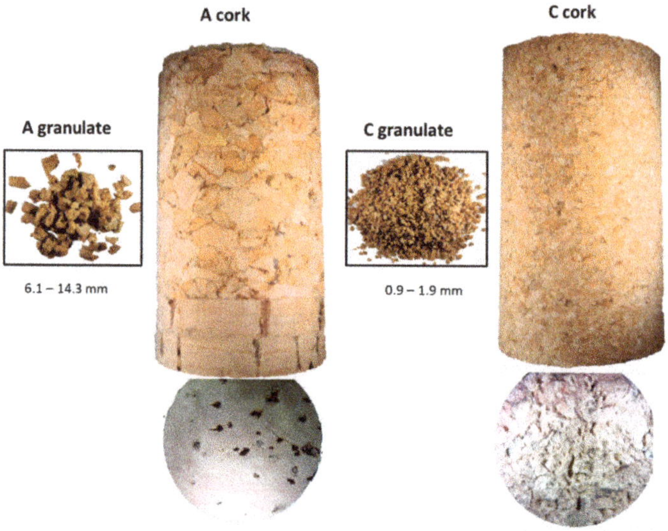

Figure 1. Representative image of the studied granulates and cork stoppers (A, high particle size and C, low particle size).

2.2. Reagents

Protocatechuic acid and gallic acid were purchased from Sigma-Aldrich (Steinheim, Germany). Folin–Ciocalteu reagent, sodium carbonate anhydrous, ethyl alcohol 96%, and acetic acid glacial were purchased from PanReac AppliChem, Barcelona, Spain. All the aromatic standards were supplied by Sigma (Sigma Aldrich, Merck Life Science, Barcelona, Spain). Vanillin 99%, guaicol 98%, eugenol 98%, benzaldehyde ≥ 99%, nonenal 97%, phenylacetaldehyde ≥ 90%, phenylethyl alcohol ≥ 98%, benzyl alcohol ≥ 99%, camphor 96%, borneol 97%, 4-terpineol ≥ 96%, α-terpineol 90%, γ-nonalactone ≥ 98%, nonanoic acid ≥ 97%, vanillic acid ≥ 97%, octanoic acid ≥ 99%, dodecanoic acid 98%, benceneacetic acid ≥ 99%, furfural 99%. Dichlorometane anhydrous 99.8%, pentane anhydrous ≥ 99%, acetonitrile anhydrous ≥ 99.8%, phosphate buffered saline (PBS), potassium peroxodisulfate ≥ 99%, 6-hidroxy-2,5,7,8-tetramethyl-chromane-2-carboxylic acid 97% (Trolox), and 2,2′-Azino-bis(3-ethylbenzothiazoline-6-sulfonic acid) diammonium salt (ABTS) were also purchased from Sigma Aldrich, Merck Life Science, Barcelona, Spain.

2.3. GC–MS Analysis

A volume of 100 µL of 2-octanol, as internal standard, was added to 100 mL of extract. Afterward, samples were separated by a SPE (solid phase extraction) cartridge (Bond Elut ENV, 500 mg and 6 mL, Agilent Tech., Santa Clara, California, USA). The cartridges were previously conditioned with 5 mL of dichloromethane, 5 mL of ethanol, and 5 mL of hydroalcoholic solution (12%). Analytes were eluted with pentane dichloromethane (50:50, *v/v*), then dried using a concentrator vacuum (Savant™ SPD131DDA, Thermo Fisher Scientific, Barcelona, Spain). Finally, samples were redissolved in 200 µL of dichloromethane.

GC analysis was performed on a GC 7890A (Agilent Tech., Santa Clara, California) system equipped with a mass spectrometer 5975C inert MSD (with Triple-Axis Detector). The column was a DB-5 (30 m × 0.25 mm × 0.25 µm, Agilent Tech.). A constant flow of 2.1 mL/min of He was used as carrier gas. Five microliters of sample was injected in splitless mode with 17.33 psi pressure (septum purge flow 15 mL/min and splitless time 1 min.). The injector temperature was maintained at 225 °C for 1 min and then heated up to 250 °C at 5 °C/min. The temperature of the oven (40 °C) was maintained for 1 min and then increased up to 260 °C at (20 °C/min.) for 25 min. The mass spectrometer operated at 70 eV (electron ionization) modes. The analysis was performed in Scan mode (m/z 10–1000). The compounds were identified by retention times and mass fragments, to compare with those of pure standard compounds. The quantification was carried out using internal standard patterns.

2.4. Total Phenolic Content

The phenolic composition was determined by the Folin–Ciocalteu assay and HPLC-DAD/MS analysis. Total phenols (TP) were determined using the Folin–Ciocalteu assay [21] with some modifications. Briefly, 100 µL of sample, 500 µL of Folin–Ciocalteu reagent, and 2 mL of a solution of sodium carbonate (20% *w/v*) were mixed, final volume 10 mL with water. The solution was stocked for 30 min for the reaction to take place and stabilize and finally, the absorbance was measured at 750 nm.

Chromatographic analyses were carried out in an Agilent 1200 series (Agilent Technologies, Palo Alto, CA, USA) coupled with DAD and MS detectors. A volume of 50 mL of each extract was washed 3 times with 20 mL of ethyl acetate. The organic phases were pooled and evaporated, re-dissolved in 1 mL of water/methanol (50:50) and then identified and quantified by HPLC–DAD/MS. A Zorbax Eclipse Plus C18 column (3.5 µm, 150 × 4.6 mm) was used. The chromatographic conditions were used according to Azevedo et al. (2014) [20]. Briefly, solvent A was 0.1% of acetic acid in water, solvent B was acetic acid, acetonitrile, and water (1:20:79, *v*/*v*/*v*). The gradient was from 80% to 20% of solvent A over 55 min, from 20% to 10% of A from 55 to 70 min, and from 10% to 0% of A from 70 to 90 min. The flow rate was 0.3 mL/min, and the sample injection 20 μL. Gallic acid and protocatechuic acid were identified by the retention time and UV–VIS spectra. The compounds were quantified with phenolic standards using peak area data of resolved peaks at 280 nm. The corresponding calibration curves were made up for gallic acid (r^2 = 0.999) and protocatechuic acid (r^2 = 0.999). The identity of the phenolic compounds was confirmed by mass spectrometry. A TSQ Quantum™ Access MAX (Thermo Fisher Scientific, Waltham, MA USA) equipped with an HESI (Heated Electrospray Ionization) source which was operated in the negative ionization mode between *m/z* 80 and 800 was used. The HESI spray voltage was set at 3.5 kV, and the capillary temperature was maintained at 350 °C. Nitrogen was used for nebulization and desolvation (sheath gas 60 arb. and auxiliary gas 20 arb.). The vaporizer temperature was maintained at 350 °C. Argon was used as the collision gas for collision-induced dissociation.

2.5. Antioxidant Activity Assay

The ABTS method allows determining the antioxidant activity through the discoloration of the cationic $ABTS^+$ radical produced by the oxidation of ABTS with potassium persulphate. This assay was performed according to the procedure described by Re. et al. (1999) [22] with slight modifications. A stock solution of 7 mM ABTS in water was prepared. To form the radical cation, a solution of potassium persulfate (2.45 mM) was prepared, using the ABTS stock solution as solvent. This solution was stored at 4 °C in the absence of light to complete the reaction. To prepare the working reagent, the solution was diluted with phosphate buffered saline (PBS) at pH 7.4, until an absorbance around 0.7 at 734 nm was obtained. Trolox or samples were added, and the decrease on the absorbance at 734 nm was measured, since the coloration disappears when the radical is reduced by antioxidants. Blank was made by adding 2 mL of reagent in a cuvette, and its absorbance was measured at 734 nm. Subsequently, 50 μL of diluted sample was added and vortexed for 30 s. After 4 min of incubation at room temperature, the absorbance was measured again at 734 nm, results were expressed in μmol/L of Trolox reagent. Samples were analyzed in several concentrations, making five dilutions of each sample by duplicate.

2.6. Statistics

Xlstat 2016.01 statistical software (Microsoft Ibèrica, Barcelona, Spain) add-on for Microsoft Excel package (Microsoft Ibèrica, Barcelona, Spain) was used for data processing. Significant differences between granulates and cork samples were determined by one-way ANOVA using the Tukey's HSD (honestly significant difference) test, at 95% of confidence level.

3. Results and Discussion

3.1. Aromatic Characterization

A total of 23 aromatic compounds were determined in granulates and cork samples and grouped according their chemical structures. Vanillins and derivatives, volatile phenols, aldehydes, alcohols, terpenes, lactones, fatty acids, and furans were found. A lot of these volatiles are closely related to certain pleasant aromatic descriptors. Table 1 shows the minimum and maximum content of the individual compounds found. The most important compound in the analyzed samples was vanillin (up to 170 μg/g), and to a lesser extent 4-vinylguaicol (23 μg/g), acetovanillone (14 μg/g), and dodecanoic acid (6.3 μg/g). All these volatiles have very pleasant aromas, such as vanilla, coconut, or wood, highly used in culinary industry and cosmetics.

Table 1. Aromatic compounds, families, descriptors, and minimum and maximum content found in the studied granulates and cork macerates.

Aromatic Compound	Aromatic Descriptor	Content (μg/g)
Vainillins		
Vanillin	vanilla	9–170
Acetovanillone	vanilla	0.6–14
Volatile phenols		
Guaicol	wood, smoked, sweet, medicine	0.03–5.0
4-vinylguaicol	wood, spice cloves, curry	0.5–23
Eugenol	spice cloves, honey	0.01–0.3
Isoeugenol	carmination	0.06–2.4
Cerulignol	spicy	0.04–2.2
Aldehydes		
Benzaldehyde	almonds, sweet, caramel	0.02–0.21
Nonenal	wax, citrus	0.03–0.47
Phenylacetaldehyde	green, grass, honey	0.05–4.5
Alcohols		
Phenylethyl alcohol	flowers, honey, pollen	0.01–2.26
Benzyl alcohol	roses, almond	0.05–0.13
Terpenols		
Camphor	mint	0–0.23
Borneol	pine tree	0–0.2
4-terpineol	spices, wood, soil	0.02–0.14
α-terpineol	flowers, lilac, sweet	0–0.2
Lactones		
γ-nonalactone	coconut, peach	0.03–0.11
Fatty acids		
Nonanoic acid	wax, dry, fatty	0.12–0.67
Vanillic acid	vanilla	0.07–0.86
Octanoic acid	coconut, lactic, rancid, cheese, sweat	0.14–3.38
Dodecanoic acid	coconut, fatty, metallic	0–6.3
Benceneacetic acid	honey, fruity, sour	0–3.0
Furans		
Furfural	caramel, candy	0–0.19

Figure 2a shows the sum of the total aromatic content of granulates and corks. In this figure, we can observe that the granulate samples extracted about 75% more than the cork stopper samples. Curiously, the highest amount of aromas were extracted in A granulate, corresponding to the largest particle size, followed by B granulate (medium size), and finally C granulate (smallest size). This could be due to the differences on the weight-volume relationship of granulates and their porosity, since the natural cork is a heterogeneous material with structural differences [23]. In this case, the smaller size of particles did not contribute to an increase of the extraction, likely because of differences in the volume of lenticels and dense matter [24]. Regarding cork stoppers, significant differences were not observed, A and B extracted a similar amount of aromas, slightly higher than that of C. Granulate samples (A and B) obtained the highest concentrations of total aromas after 5 days of maceration (Figure 2b), while for corks the maximum content was generally obtained at the end of the assay (after 15 days of extraction). In granulate samples, the highest amount of aromatic compounds may be reached earlier than in cork samples, likely because the characteristics of the samples and size facilitate the extraction of volatiles. In the studied samples, it seems that the bigger the granulates, the faster the extraction rate of the volatiles. The A granulates that correspond with the highest granulate size, mean value around 9.2 mm, obtained the highest amount of volatiles reaching levels over 200 µg/g. This increment in A granulate at 5 days is mainly due to a high extraction of vanillins and volatile phenols, and to a lesser extent to terpenols and fatty acids (Figure 3).

Considering the time of extraction in cork stoppers, only in the case of A corks was the maximum value of volatiles reached in 5 days of maceration, since for B and C corks, the maximum values were reached after 15 days of maceration (Figure 2).

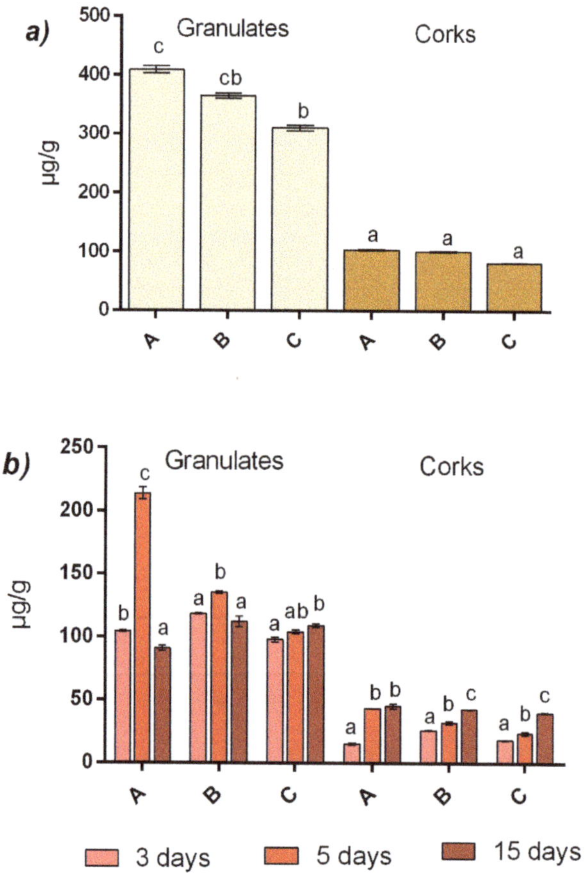

Figure 2. Total aromatic compounds (mean ± SD) of granulates and corks (**a**) and total aromatic compounds of granulates and corks at different time of extraction (**b**). Different particle size from cork stoppers and their respective granulates (A; high, B; medium, C; small). Different letters indicate significant differences, $p < 0.05$.

Figure 3 shows the total aromatic composition of the extracts per studied family. Vanillins are the most important family. In this study, this compound reached values from 9.3 µg/g (in C corks) to 167.9 µg/g (in A granulates, Table S2). Therefore, vanillin has a great impact on the overall aroma of cork extracts. This compound is very characteristic for providing a very intense and pleasant aroma with sweet and floral notes, which is currently used in many different industrial fields, such as in the production of fragrances, in food use as in baking [25], or in cosmetics [11]. In addition, as was seen in other studies [26,27], vanillin is a natural bioactive component present in corks, which showed an

important antioxidant activity. Another important aromatic family of compounds studied in the food industry were volatile phenols [28,29]. As shown in Figure 3, the maximum content of these aromatic compounds was extracted after 5 days of maceration in A and B granulates. This pattern seems to be also reproduced in A and B cork samples. It may be due to the high concentration of vinylguaiacol (values up to 23 μg/g) and, to a lesser extent, to guaiacol content (5 μg/g). Both compounds are characteristic for having aromas of spicy notes, specifically clove, wood, and smoked [30,31].

Regarding aldehydes, it was observed that the granulates extracted the highest concentrations after 15 days of maceration with values up to five times higher with respect to the corks. The main compounds were phenylacetaldehyde with maximum values of 4.5 μg/g and nonenal with values of 0.47 μg/g. These compounds are characteristic for providing fresh and intense aromas even at low concentrations, such as green grass, citrus, and wax [32]. It should be noted that the odor threshold for phenylacetaldehyde in hydroalcoholic solution was established in 5 μg/L [33], and the maximum content found in this work corresponded to 135 μg/L (30 g of cork in 1 L). In this way, the maximum nonenal amount in these samples was 14.1 μg/L higher than the established odor threshold determined in water (0.065 μg/L) and in hydroalcoholic solution (0.17 μg/L) [34].

Lactones exhibited a behavior similar to that of aldehydes. The aroma of lactones is of interest for commercial aromatization of food [35]. Here, γ-nonalactone is the compound that represents this family. This compound is better extracted from granulates than from corks after 15 days of maceration. Despite presenting low concentrations (0.11 μg/g) in the extracts, its concentration is higher than the odor threshold in water (0.03 μg/g) [35]. This is a characteristic compound for providing pleasant aromatic descriptors, such as coconut, peach, and sweet cream butter notes [36,37]. This aroma is commonly used in the development of cosmetics and fragrances. In addition, they could also be used in the food industry as an essence in the preparation of cakes, sweets, candies, and ice cream, as margarine or aroma [36].

Similarly, the terpenols had a higher concentration in the macerates of granulates than in corks, especially after 15 days of extraction. The contribution of the individual terpenols seems to be similar, since little differences were observed among them. Their concentrations ranged from 0.05 (A cork after 3 days) to 0.62 μg/g (A granulate after 5 days) (Table S2). Terpenes are aromatic compounds commonly synthesized in plants, trees, and vegetables. They are usually the main constituents of essential oils of most plants, offering a wide variety of pleasant scents, from flowery to fruity, to woody, or balsamic notes [38]. Hence, cork has a wide range of terpenoid variety within the family of terpenols formed in this study by the following determined compounds: camphor, borneol, α-terpineol, and 4-terpineol. All of them have interesting aromatic descriptors such as mint, pine, spices, and flowers, respectively. In addition, today they are used in cosmetics, especially in the elaboration of anti-aging creams, because they possess bioactive properties [39]. They may act as elastase inhibitors, preventing the structural degradation of elastin fibers in the dermal matrix [2,40,41], and play an important role as constituents of flavors for spicing foods, sweets, beverages, and baked foods [42]. Furfural, an important aromatic additive in food and beverages [43], also showed the highest concentrations after 15 days of maceration, being higher in granulates (5.7 μg/L) than in corks (1.5 μg/L). The odor threshold of this compounds is also lower (1 μg/L) [44] than the concentrations found in the studied samples, as occurred in other compounds.

On the other hand, alcohols and fatty acids showed higher concentrations in macerated cork stoppers than in granulates (Figure 3). With regard to alcohols, granulates and corks showed a large increase in concentration after 15 days of maceration, reaching values of up to 0.98 μg/g and 2.32 μg/g, respectively. Among them, the phenylethyl alcohol stands out for presenting the highest concentrations (Table S2), and for having interesting aromatic descriptors, such as flowers. For this reason, this compound is commonly used in the production of perfumes as well as flavoring in the food industry. It is also used in the cosmetic industry for the preparation of creams and soaps, acting as a preservative due to its stability in basic conditions. Another important property of this compound is its antimicrobial activity [45]. Fatty acids showed a large increase in concentration after 5 and 15 days of maceration in cork stoppers. This family reached a value of 10.74 µg/g. It should be noted that their concentrations were three times higher than in the case of granulates. The main fatty acids determined were vanillic, bezenacetic, decanoic, and octanoic, highlighting the latter for having the highest concentrations. All of them contributed to providing very pleasant aromatic notes such as vanilla, honey, lactic, and coconut, respectively.

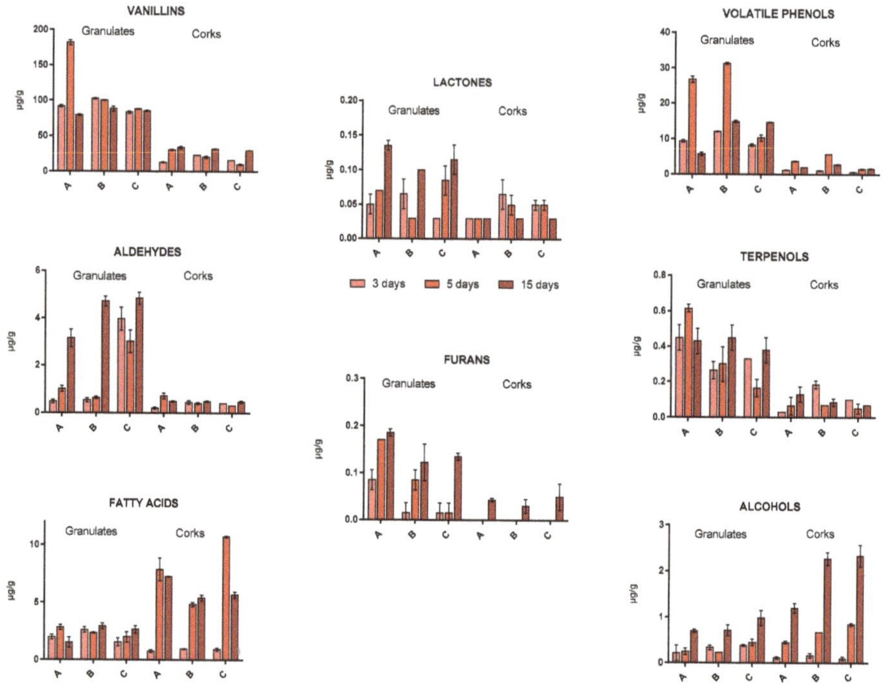

Figure 3. Composition per family of the studied aromatic compounds. Different particle size from cork stoppers and their respective granulates (A; high, B; medium, C; small).

The great increase in the concentration of phenylethyl alcohol and octanoic acid in the macerates of corks with respect to granulates could be due to the presence of this compound in glues used for the manufacture of agglomerated corks [46]. Their use is due to their aromatic, antimicrobial, and antibiotic properties.

In summary, the families studied provide the aromatic profile of the granules and cork stoppers, which are rich in very pleasant aromas at the sensory level. This aromatic composition extracted from cork could have a second shelf life in different types of industries. On the one hand, the cosmetic and pharmaceutical industries could use these extracted compounds as ingredients in the manufacture of products such as sunscreens, wrinkle products, fragrances, or even soaps. On the other, they could be also used in processed food as flavoring additive, and it should be noted their bioactive properties and beneficial health effects, such as antioxidant and antimicrobial activities.

3.2. Phenolic Composition

Many studies have used the Folin–Ciocalteu method to determine polyphenols in plant extracts [47,48]. The combination of HPLC–DAD/MS analysis with this methodology helps the determination of the compounds, from the individual to the polymeric polyphenols. Figure 4 shows the Folin–Ciocalteu index (Figure 4a,b) and the content in the phenolic compounds in the studied cork samples determined by HPLC (Figure 4c,d).

In general, higher polyphenol content was obtained for granulates than for corks. A and B granulates obtained the highest value after three days of maceration (Figure 4a). As shown in Figure 4c,d, in general, a higher concentration of phenolics was extracted after 15 days of maceration. In the case of granulates, it is observed that A and B granulates had concentrations above 150 µg/g after 3 and 5 days of maceration, and increased to concentrations around 500 µg/g after 15 days. The highest concentration was obtained in A granulate (513.5 µg/g), with larger particle size. In C granulate (smaller size), the extraction was more constant, around 350–400 µg/g for all maceration times. After 15 days of maceration, the phenolic composition in cork stoppers increased in all cases from 10 µg/g to 65 µg/gin the C sample. Other authors observed differences in the phenolic extraction depending on the type of cork stopper, granulate, and powder [12,20,49]. Further research should be done to optimize time of extraction and methods. Furthermore, the use of new promising technologies may be interesting in order to optimize the process in a real scale [50].

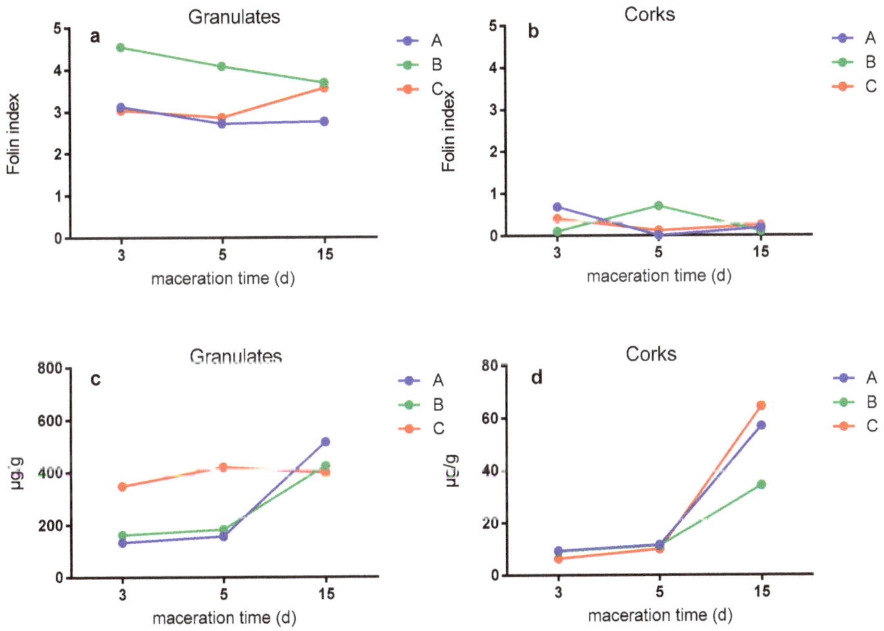

Figure 4. Folin–Ciocalteu index (**a** and **b**) and phenolic content determined by HPLC (**c** and **d**; note difference in y-axis scales). Particle size from cork stoppers and their respective granulates, A; high, B; medium, C; small.

Considering the individual content of the phenolic compounds, gallic acid obtained the highest concentrations with values between 60.6 and 180.9 µg/g, followed by protocatechuic acid with values between 41.3 and 161.6 µg/g, and finally the protocatechuic aldehyde with values between 26.3 and 118.6 µg/g. As stated in previous studies [51,52], gallic acid and protocatechuic acid are phenolic compounds with abundant presence in cork extractive fractions.

3.3. Antioxidant Activity

Previous studies have already shown that the cork has bioactive compounds with antioxidant activity [2,53]. Touati et al. studied cork extracts in methanol solution and water using the ABTS method, demonstrating its high antioxidant activity, and Fernandes et al. characterized the great antioxidant activity of cork extracts, which was directly related to the phenolic composition [15,54].

Figure 5 shows the antioxidant activity in μmol/L of Trolox of the granules and corks throughout the maceration time. In general, it is observed that granulate samples obtained higher antioxidant activity than corks (up to one hundred times). Figure 5A shows that B and C granulates had higher antioxidant activity compared with the larger A granulate. Likewise, the three types of granulates obtained similar antioxidant activity after all maceration times, as reported Azevedo and co-workers, who observed no significant differences in wine model solution at different times when bottling with different cork stoppers [20]. With respect to the corks (Figure 5B), in general, the same trend is observed in all types of corks. In this case, the antioxidant activity increases by increasing the maceration time. The differences observed between samples after 15 days of maceration may result from the higher amount of simpler phenolic compounds at this time (Figure 4d). The maceration of A cork (larger granulate) presented the highest concentration at all times, especially after 15 days of maceration. On the one hand, this trend is not explained from the obtained data, so this could be due to hydrolyzable tannins that are powerful antioxidant agents and were not determined in this study [15].

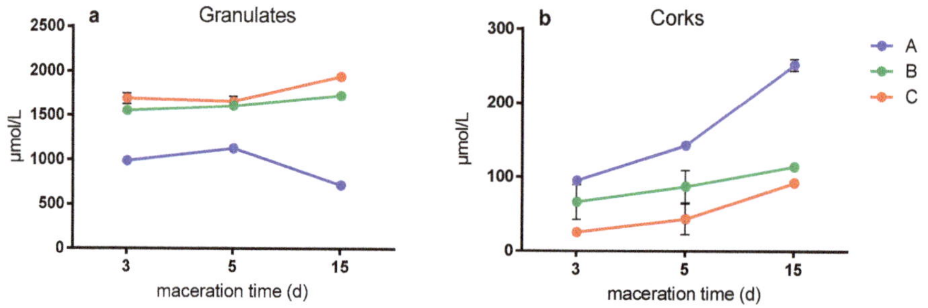

Figure 5. Antioxidant activity of granulates (**a**) and cork (**b**) samples determined by ABTS method (expressed in μmol/L of Trolox reagent), note difference in *y*-axis scales. Particle size from cork stoppers and their respective granulates, A; high, B; medium, C; small.

On the other hand, the high antioxidant capacity of the cork samples studied, especially in the case of granulates, could be directly related to phenolic acids, in particular the high concentrations of gallic acid and protocacatechuic acid [26,54]. In addition, this antioxidant capacity could also be due to the aromatic composition of the corks studied, especially to the high concentrations of vanillin (170 μg/g), which is considered a natural bioactive component of cork [15,55]. According to Azevedo et al., the amounts of gallic acid, protocatechuic acid, protocatechuic aldehyde, and vanillin in cork samples appear to be crucial for the antioxidant activity [20].

4. Conclusions

Overall, this work highlights valuable aromatic compounds (vanillins, volatile phenols, aldehydes, alcohols, terpenols, lactones, fatty acids, and furans) found in cork by-products and cork stoppers to be reused as flavoring agents and antioxidants in the food industry. The most important family of compounds was vanillins with a high content of vanillin and acetovanillone, and 4-vinylguaicol was the most abundant volatile phenol. Furthermore, the phenolic composition of corks and their by-products may provide interesting antioxidant properties with increasing interest on the industry for their health benefits. Granulates showed higher potential than corks, however, corks used daily (e.g., wine stoppers) have to be considered for this purpose. Further research should be done in order to optimize waste management and extraction procedures, among others.

Supplementary Materials: The following are available online at http://www.mdpi.com/2304-8158/9/2/133/s1. Table S1: Size of cork granulates (mm) used in this study (n = 20), Table S2: Individual aromatic composition in the studied granulates and corks at 3, 5, and 15 days of maceration.

Author Contributions: A.M.M. performed the experiment and participated on the manuscript redaction and edition. M.P. designed the experiment and reviewed the manuscript. R.F.-G. participated on the experiment analysis, redaction and edition. All authors have read and agreed to the published version of the manuscript.

Acknowledgments: Authors thanks the European project entitled "Efficiency in the use of resources for the improvement of the wine sustainability in Priorat region" Life Priorat+Montsant, Ref. LIFE15ENV/ES/000399.

Conflicts of Interest: The authors declare no conflict of interest

References

1. Gil, L. Cork powder waste: An overview. *Biomass Bioenerg.* **1997**, *13*, 59–61. [CrossRef]
2. Carriço, C.; Ribeiro, H.M.; Marto, J. Converting cork by-products to ecofriendly cork bioactive ingredients: Novel pharmaceutical and cosmetics applications. *Ind. Crop. Prod.* **2018**, *125*, 72–84. [CrossRef]
3. Borrero, C.; Castillo, S.; Casanova, E.; Segarra, G.; Trillas, M.I.; Castaño, R.; Avilés, M. Capacity of composts made from agriculture industry residues to suppress different plant diseases. *Act. Hortic.* **2013**, *1013*, 459–464. [CrossRef]
4. Carmona, E.; Ordovás, J.; Moreno, M.T.; Avilés, M.; Aguado, M.T.; Ortega, M.C. Hydrological Properties of Cork Container Media. *HortScience* **2003**, *38*, 1235–1241. [CrossRef]
5. Filipe-Ribeiro, L.; Cosme, F.; Nunes, F.M. A Simple Method To Improve Cork Powder Waste Adsorption Properties: Valorization as a New Sustainable Wine Fining Agent. *ACS Sustain. Chem. Eng.* **2019**, *7*, 1105–1112. [CrossRef]
6. Silva, B.; Martins, M.; Rosca, M.; Rocha, V.; Lago, A.; Neves, I.C.; Tavares, T. Waste-based biosorbents as cost-effective alternatives to commercial adsorbents for the retention of fluoxetine from water. *Sep. Purif. Technol.* **2020**, *235*, 116139. [CrossRef]
7. Pintor, A.M.A.; Vieira, B.R.C.; Boaventura, R.A.R.; Botelho, C.M.S. Removal of antimony from water by iron-coated cork granulates. *Sep. Purif. Technol.* **2020**, *233*, 116020. [CrossRef]
8. Pirozzi, C.; Pontoni, L.; Fabbricino, M.; Bogush, A.; Campos, L.C. Effect of organic matter release from natural cork used on bisphenol a removal from aqueous solution. *J. Clean. Prod.* **2020**, *244*, 118675. [CrossRef]
9. Olivella, M.À.; Jové, P.; Oliveras, A. The use of cork waste as a biosorbent for persistent organic pollutants-Study of adsorption/desorption of polycyclic aromatic hydrocarbons. *J. Environ. Sci. Health. Part A* **2011**, *46*, 824–832. [CrossRef]
10. OIV. *2019 Statistical Report on World Vitiviniculture*; International Organization of Vine and Wine: Paris, France, 2019; p. 13.
11. Aroso, I.M.; Araújo, A.R.; Pires, R.A.; Reis, R.L. Cork: Current Technological Developments and Future Perspectives for this Natural, Renewable, and Sustainable Material. *ACS Sustain. Chem. Eng.* **2017**, *5*, 11130–11146. [CrossRef]
12. Reis, S.F.; Lopes, P.; Roseira, I.; Cabral, M.; Mateus, N.; Freitas, V. Recovery of added value compounds from cork industry by-products. *Ind. Crop. Prod.* **2019**, *140*, 111599. [CrossRef]
13. Mark, R.; Lyu, X.; Lee, J.J.L.; Parra-Saldívar, R.; Chen, W.N. Sustainable production of natural phenolics for functional food applications. *J. Funct. Foods* **2019**, *57*, 233–254. [CrossRef]
14. Santos-Buelga, C.; Scalbert, A. Proanthocyanidins and tannin-like compounds - Nature, occurrence, dietary intake and effects on nutrition and health. *J. Sci. Food Agric.* **2000**, *80*, 1094–1117. [CrossRef]
15. Fernandes, A.; Fernandes, I.; Cruz, L.; Mateus, N.; Cabral, M.; de Freitas, V. Antioxidant and Biological Properties of Bioactive Phenolic Compounds from Quercus suber L. *J. Agric. Food. Chem.* **2009**, *57*, 11154–11160. [CrossRef] [PubMed]
16. Fontana, A.R. Analytical methods for determination of cork-taint compounds in wine. *TrAC, Trends Anal. Chem.* **2012**, *37*, 135–147. [CrossRef]
17. Garcia, A.R.; Lopes, L.F.; Barros, R.B.D.; Ilharco, L.M. The problem of 2,4,6-trichloroanisole in cork planks studied by attenuated total reflection infrared spectroscopy: Proof of concept. *J. Agric. Food. Chem.* **2015**, *63*, 128–135. [CrossRef]
18. Mayr, C.M.; Capone, D.L.; Pardon, K.H.; Black, C.A.; Pomeroy, D.; Francis, I.L. Quantitative Analysis by GC-MS/MS of 18 Aroma Compounds Related to Oxidative Off-Flavor in Wines. *J. Agric. Food. Chem.* **2015**, *63*, 3394–3401. [CrossRef]

19. Tarasov, A.; Rauhut, D.; Jung, R. "Cork taint" responsible compounds. Determination of haloanisoles and halophenols in cork matrix: A review. *Talanta* **2017**, *175*, 82–92. [CrossRef]
20. Azevedo, J.; Fernandes, I.; Lopes, P.; Roseira, I.; Cabral, M.; Mateus, N.; Freitas, V. Migration of phenolic compounds from different cork stoppers to wine model solutions: antioxidant and biological relevance. *Eur. Food Res. Technol.* **2014**, *239*, 951–960. [CrossRef]
21. OIV. *Compendium of International Methods of Analysis*; International Organization of Vine and Wine: Paris, France, 2019.
22. Re, R.; Pellegrini, N.; Proteggente, A.; Pannala, A.; Yang, M.; Rice-Evans, C. Antioxidant activity applying an improved ABTS radical cation decolorization assay. *Free Radical Bio. Med.* **1999**, *26*, 1231–1237. [CrossRef]
23. Oliveira, V.; Knapic, S.; Pereira, H. Natural variability of surface porosity of wine cork stoppers of different commercial classes. *J. Int. Sci. Vigne Vin.* **2012**, *46*, 331–340. [CrossRef]
24. Crouvisier-Urion, K.; Chanut, J.; Lagorce, A.; Winckler, P.; Wang, Z.; Verboven, P.; Nicolai, B.; Lherminier, J.; Ferret, E.; Gougeon, R.D.; et al. Four hundred years of cork imaging: New advances in the characterization of the cork structure. *Sci. Rep.* **2019**, *9*, 19682. [CrossRef] [PubMed]
25. Ciriminna, R.; Fidalgo, A.; Meneguzzo, F.; Parrino, F.; Ilharco, L.M.; Pagliaro, M. Vanillin: The Case for Greener Production Driven by Sustainability Megatrend. *ChemistryOpen* **2019**, *8*, 660–667. [CrossRef] [PubMed]
26. Rodrigues, F.; Pimentel, F.B.; Oliveira, M.B.P.P. Olive by-products: Challenge application in cosmetic industry. *Ind. Crop. Prod.* **2015**, *70*, 116–124. [CrossRef]
27. Silva, S.P.; Sabino, M.A.; Fernandes, E.M.; Correlo, V.M.; Boesel, L.F.; Reis, R.L. Cork: properties, capabilities and applications. *Int. Mater. Rev.* **2005**, *50*, 345–365. [CrossRef]
28. Salgado, J.M.; Rodríguez-Solana, R.; Curiel, J.A.; Rivas, B.d.l.; Muñoz, R.; Domínguez, J.M. Production of vinyl derivatives from alkaline hydrolysates of corn cobs by recombinant Escherichia coli containing the phenolic acid decarboxylase from Lactobacillus plantarum CECT 748T. *Bioresour. Technol.* **2012**, *117*, 274–285. [CrossRef]
29. Sharma, U.K.; Sharma, N.; Salwan, R.; Kumar, R.; Kasana, R.C.; Sinha, A.K. Efficient synthesis of hydroxystyrenes via biocatalytic decarboxylation/deacetylation of substituted cinnamic acids by newly isolated Pantoea agglomerans strains. *J. Sci. Food Agric.* **2012**, *92*, 610–617. [CrossRef]
30. Culleré, L.; Cacho, J.; Ferreira, V. Comparative study of the aromatic profile of different kinds of wine cork stoppers. *Food Chem.* **2009**, *112*, 381–387. [CrossRef]
31. Francis, I.L.; Newton, J.L. Determining wine aroma from compositional data. *Aus. J. Grape Wine R.* **2005**, *11*, 114–126. [CrossRef]
32. Rocha, S.; Delgadillo, I.; Ferrer Correia, A.J. GC–MS Study of Volatiles of Normal and Microbiologically Attacked Cork from Quercus suber L. *J. Agric. Food. Chem.* **1996**, *44*, 865–871. [CrossRef]
33. Ferreira, V.; López, R.; Cacho, J.F. Quantitative determination of the odorants of young red wines from different grape varieties. *J. Sci. Food Agric.* **2000**, *80*, 1659–1667. [CrossRef]
34. Chatonnet, P.; Dubourdieu, D. Identification of substances responsible for the 'sawdust' aroma in oak wood. *J. Sci. Food Agric.* **1998**, *76*, 179–188. [CrossRef]
35. Belitz, H.-D.; Grosch, W.; Schieberle, P. *Food Chemistry*, 4th ed.; Springer: Berlin Heidelberg, Germany, 2009.
36. Dadalı, C.; Elmacı, Y. Optimization of Headspace-Solid Phase Microextraction (HS-SPME) technique for the analysis of volatile compounds of margarine. *J. Food Sci. Technol.* **2019**, *56*, 4834–4843. [CrossRef] [PubMed]
37. Mosandl, A.; Günther, C. Stereoisomeric Flavor Compounds. 20. Structure and Properties of γ-Lactone Enantiomers. *J. Agric. Food. Chem.* **1989**, *37*, 413–418. [CrossRef]
38. Lorenzo, C.; Eugenio, A. Use of Terpenoids as Natural Flavouring Compounds in Food Industry. *Recent Pat. Food. Nutr.* **2011**, *3*, 9–16.
39. Brahmkshatriya, P.P.; Brahmkshatriya, P.S. Terpenes: Chemistry, biological role, and therapeutic applications. In *Natural Products: Phytochemistry, Botany and Metabolism of Alkaloids, Phenolics and Terpenes*; Springer: Berlin/Heidelberg, Germany, 2013; pp. 2665–2691.
40. Karim, A.A.; Azlan, A.; Ismail, A.; Hashim, P.; Abd Gani, S.S.; Zainudin, B.H.; Abdullah, N.A. Phenolic composition, antioxidant, anti-wrinkles and tyrosinase inhibitory activities of cocoa pod extract. *BMC Complem. Altern. M.* **2014**, *14*, 381.
41. Mukherjee, P.K.; Maity, N.; Nema, N.K.; Sarkar, B.K. Bioactive compounds from natural resources against skin aging. *Phytomedicine* **2011**, *19*, 64–73, Eberhard. [CrossRef]

42. Breitmaier, E. *Terpenes: Flavors, Fragrances, Pharmaca, Pheromones*; Wiley-VCH Verlag GmbH & Co. KGaA: Weinheim, Germany, 2006; pp. 1–214.
43. Hidalgo-Carrillo, J.; Marinas, A.; Urbano, F.J. Chemistry of Furfural and Furanic Derivatives. In *Furfural*; World Scientific (Europe): London, UK, 2017; Volume 2, pp. 1–30.
44. Gómez-Míguez, M.J.; Cacho, J.F.; Ferreira, V.; Vicario, I.M.; Heredia, F.J. Volatile components of Zalema white wines. *Food Chem.* **2007**, *100*, 1464–1473. [CrossRef]
45. Ren, G.; Xue, P.; Sun, X.; Zhao, G. Determination of the volatile and polyphenol constituents and the antimicrobial, antioxidant, and tyrosinase inhibitory activities of the bioactive compounds from the by-product of Rosa rugosa Thunb. var. plena Regal tea. *BMC Complem. Altern. M.* **2018**, *18*, 307. [CrossRef]
46. Crouvisier-Urion, K.; Bellat, J.P.; Gougeon, R.D.; Karbowiak, T. Mechanical properties of agglomerated cork stoppers for sparkling wines: Influence of adhesive and cork particle size. *Compos. Struct.* **2018**, *203*, 789–796. [CrossRef]
47. Blainski, A.; Lopes, G.C.; De Mello, J.C.P. Application and analysis of the folin ciocalteu method for the determination of the total phenolic content from limonium brasiliense L. *Molecules* **2013**, *18*, 6852–6865. [CrossRef] [PubMed]
48. Cicco, N.; Lanorte, M.T.; Paraggio, M.; Viggiano, M.; Lattanzio, V. A reproducible, rapid and inexpensive Folin-Ciocalteu micro-method in determining phenolics of plant methanol extracts. *Microchem. J.* **2009**, *91*, 107–110. [CrossRef]
49. Fernandes, A.; Sousa, A.; Mateus, N.; Cabral, M.; de Freitas, V. Analysis of phenolic compounds in cork from Quercus suber L. by HPLC–DAD/ESI–MS. *Food Chem.* **2011**, *125*, 1398–1405. [CrossRef]
50. Albanese, L.; Bonetti, A.; D'Acqui, L.P.; Meneguzzo, F.; Zabini, F. Affordable production of antioxidant aqueous solutions by hydrodynamic cavitation processing of silver fir (*Abies alba* Mill.) needles. *Foods* **2019**, *8*, 65. [CrossRef] [PubMed]
51. Santos, S.A.O.; Pinto, P.C.R.O.; Silvestre, A.J.D.; Neto, C.P. Chemical composition and antioxidant activity of phenolic extracts of cork from *Quercus suber* L. *Ind. Crop. Prod.* **2010**, *31*, 521–526. [CrossRef]
52. Santos, S.A.O.; Villaverde, J.J.; Sousa, A.F.; Coelho, J.F.J.; Neto, C.P.; Silvestre, A.J.D. Phenolic composition and antioxidant activity of industrial cork by-products. *Ind. Crop. Prod.* **2013**, *47*, 262–269. [CrossRef]
53. Igueld, S.B.; Hatem, A.; Trabelsi-Ayadi, M.; Cherif, J.K. Study of Physicochemicals Characteristics and Antioxidant Capacity of Cork Oak Acorns (*Quercus Suber* L.) grown in Three Regions in Tunisia. *J. App. Pharm. Sci.* **2015**, *5*, 26–32. [CrossRef]
54. Touati, R.; Santos, S.A.O.; Rocha, S.M.; Belhamel, K.; Silvestre, A.J.D. The potential of cork from Quercus suber L. grown in Algeria as a source of bioactive lipophilic and phenolic compounds. *Ind. Crop. Prod.* **2015**, *76*, 936–945. [CrossRef]
55. Anand, A.; Khurana, R.; Wahal, N.; Mahajan, S.; Mehta, M.; Satija, S.; Sharma, N.; Vyas, M.; Khurana, N. Vanillin: A comprehensive review of pharmacological activities. *Plant Arch.* **2019**, *19*, 1000–1004.

foods

MDPI

Article

Reutilization of Food Waste: One-Step Extration, Purification and Characterization of Ovalbumin from Salted Egg White by Aqueous Two-Phase Flotation

Bin Jiang, Jiaxin Na, Lele Wang, Dongmei Li, Chunhong Liu and Zhibiao Feng *

Department of Applied Chemistry, Northeast Agricultural University, NO. 600 Changjiang Road Xiangfang District, Harbin 150030, China

* Correspondence: fengzhibiao@neau.edu.cn; Tel.: +86-451-5519-0222

Received: 25 June 2019; Accepted: 23 July 2019; Published: 25 July 2019

Abstract: For the purpose of reducing pollution and the reutilization of salted egg whites, which are byproducts of the manufacturing process of salted egg yolks and normally treated as waste, an aqueous two-phase flotation (ATPF) composed of polyethylene glycols (PEG 1000) and $(NH_4)_2SO_4$ was applied to develop a simple, inexpensive and efficient process for the separation of ovalbumin (OVA) from salted egg whites. The effects of the concentration of PEG, the concentration of $(NH_4)_2SO_4$, the flow rate and the flotation time on the flotation efficiency (Y) and purity (P) of OVA were investigated. A response surface method (RSM) experiment was carried out on the basis of a single-factor experiment. An efficient separation was achieved using ATPF containing 5 mL of 80% PEG 1000 (w/w), 28 mL of 28% $(NH_4)_2SO_4$ (w/w), 35 mL/min of the flow rate and 30 min of the flotation time, while 2 mL of the salted egg white solution (salted eggs white (v): water (v) = 1:4) was loaded. Under the optimal conditions, Y and P of OVA could reach 82.15 ± 0.24% and 92.98 ± 0.68%, respectively. The purified OVA was characterized by sodium dodecyl sulfate polyacrylamide gel electrophoresis (SDS-PAGE), reverse phase high-performance liquid chromatography (RP-HPLC), liquid chromatography-nano electrospray ionisation mass spectrometry (Nano LC-ESI-MS/MS), ultraviolet spectrum (UV), fluorescence spectrum (FL) and fourier transform infrared spectroscopy (FT-IR). The results indicated that the purity of OVA obtained by ATPF was satisfactory and there was no obvious difference in the structure of the OVA separated by ATPF and the standard. The results of the functional properties revealed no significant differences between OVA obtained by ATPF and the standard in oil binding capacity, viscosity, emulsibility and foam capacity.

Keywords: reutilization of food waste; salted egg white; ovalbumin; extraction; aqueous two-phase flotation

1. Introduction

Millions of tons of losses and waste produced in the food processing industries every year cause economic, environmental, and nutritional problems. The effective utilization of these by-products, by translating waste and byproducts into resources through shifts in technology, ensures the sustainable development of food industries and a reduction of their environmental pollution.

Salted egg whites, a by-product in the manufacturing process of salted egg yolks, are usually discarded as waste due to the difficulty in treatment. Abandoned salted egg whites contain a high content of salt and organic, which is commonly regarded as an environmental problem whose disposal is troublesome for the food industry [1,2]. Salted egg white protein is a mixture of proteins that consists mainly of ovalbumin (OVA), lysozyme, ovotransferrin, ovomucoid [3]. OVA is the major protein in salted egg whites, accounting for approximately 54% of the total protein in salted egg whites. Containing a well-balanced amino acid composition, OVA is an excellent source of essential amino acids. It consists of 385 amino acids (45 kDa) with an isoelectric point (pI) of approximately 4.5 [4,5]. Although being well studied, the function of ovalbumin in eggs is still unknown. However, some studies suggested ovalbumin to be a storage protein [6]. In view of nutrition and function, OVA is one of the highest quality food proteins. Further, OVA has been proved to have biological activities, such as antibacterial activity, antihypertensive activity and immunomodulating activity [7]. In addition to displaying several functional properties including emulsification, heat-setting and foaming, OVA also has been used as an effective drug carrier [8]. Therefore, there is considerable technical interest in its separation if costs can be controlled. The current methods for separating OVA from egg white are mainly isoelectric precipitation [9,10], membrane [11] and chromatography [12]. In recent years, a few new materials were applied to separate OVA with good selectivity. However, aniline-doped cobalt mono-substituted silicotungstic acid was reported to separate OVA with an adsorption efficiency of 92% for OVA at pH 9 [13]. Liu [14] prepared a nano-composite based on reduced graphene oxide nanosheets functionalized with polymeric ionic liquid. It was used to selectively isolate OVA from egg white with an adsorption capacity of 917.4 mg g^{-1}. Chen [15] described a three-dimensional reduced graphene oxide aerogel with embedded nickel oxide nanoparticles which was prepared by a one-step self-assembly reaction, which was proved to selectively isolate ovalbumin from chicken egg white. Nevertheless, the methods above are difficult to be used in the industry due to their low yield, expensive operating costs or high equipment requirements. In addition, most previous research focused on the separation of OVA from unsalted egg white solution. Extracting OVA from salted egg whites can reprocess and utilize the waste and byproducts in the production of new commercially valuable products, as well as support an abundant and cheap source of bioactive compounds. Therefore, it is necessary to develop an economical and simple method to separate OVA from salted egg whites.

The model of the unidirectional rising bubbles mass transfer was applied to an aqueous two-phase system (ATPS), which was called aqueous two-phase flotation (ATPF). Compared with aqueous two-phase extraction (ATPE), ATPF was considered as a novel technique with low consumption of organic solvents and high concentration coefficients [16]. Thus far, ATPF has been applied in the enrichment and separation of some biomolecules, such as protein and Chinese herbal medicines. Sankaran et al. [17] established a 1-propanol/$(NH_4)_2SO_4$ ATPF to separate lipase from fermentation broth. Jiang et al. [18] used ATPF containing PEG 1000 and citrate to extract α-lactalbumin from whey. Under optimal conditions, the purification factor, purity and flotation efficiency were 5.33 ± 0.05, 96.78 ± 0.79% and 87.54 ± 0.76%, respectively. The use of ATPF to extract proteins could save process time and cost, and the use of biodegradable salts as the bottom phase could avoid environmental pollution problems. Pakhale et al. [19] used ATPF consisting of PEG and phosphate to isolate bromelain. Under optimal conditions, the purification ratio reached 4.26, and the recovery rate was 91.47%. Jiang et al. [20] used ATPF to separate antioxidant peptides from whey protein hydrolysate. The recovery rate of peptides from the top phase was 11.71% and the free radical scavenging per unit concentration was 29.18%. The result was significantly better than ATPE. Bi et al. [21] established an effective and non-polluting ATPF which isolated and enriched baicalin from Astragalus membranaceus extract. The experimental results showed that under optimal conditions of ATPF, the above method was successfully applied to the separation of baicalin, the enrichment ratio was more than 27, and the separation efficiency was 90%. Chang et al. [22] established a method for separating four flavonoids using ATPF. Under optimal conditions, the flotation efficiency was 96.03%, 92.07%, 94.27% and 90.51%, respectively. ATPF proved to be a potentially effective method for the separation of biomolecules, therefore, more and more attempts on the application of ATPF were carried out.

The aim of this study was to develop an inexpensive, simple, efficient separation and pollution-free process for the separation of OVA from salted egg whites to realize the reutilization of food waste while minimizing the environmental impact. With this goal, the separation of OVA from salted egg whites by ATPF of PEG 1000/ammonium sulphate was studied for the first time.

2. Materials and Methods

2.1. Instruments

The Waters e2695 liquid chromatograph equipped with 2998 PDA detector, C_8 and C_{18} columns (Waters, Waters Corporation, Milford, MA, USA) was applied to detect the samples. The solution pH was measured by a PHS-3C pH meter (Shanghai Instrument Electric Scientific Instrument Co., Ltd. Shanghai, China). An AL-04 electronic analytical balance (Mettler Toledo Instruments Co., Ltd., Shanghai, China) was used for weighing. The sample was dialyzed with a 13000 Da MWCO filter (Spectrum Labs, Los Angeles, CA, USA). The glass rotameter (Shenyang Beixing flow meter factory, Shenyang, China) and buoy type oxygen inhalator (Shandong Huachen high pressure container Co., Ltd. Jinan, China) were used in the preparation of ATPF.

2.2. Materials

Salted eggs were purchased from the local supermarket. The OVA standard sample was purchased from Sigma (Burlington, MA, USA). Trifluoroacetic acid (TFA) and acetonitrile were chromatographically pure (Dikma, Beijing, China). Further, 1000 $g \cdot mol^{-1}$ polyethylene glycol (PEG 1000) was purchased from Aladdin (Shanghai, China) and ammonium sulphate and other reagents were analytical reagent grade.

2.3. Preparation of ATPF and Purification

The ATPF was prepared in a flotation column by mixing 2 mL of the salted egg white solution (salted egg white (v): water (v) = 1:4) and appropriate volume of stock solution of ammonium sulfate. The gas flow outputted from the nitrogen cylinder formed a steady flow by a gas buffer, which was adjusted by the flowrator. When the nitrogen flow rate was stable, 5 mL PEG was added. The nitrogen gas was bubbling through the bottom of the column at a certain flotation time and flow velocity. The flotation unit is described in Figure 1. The OVA was carried to the PEG phase by nitrogen gas. By measuring the Y and P in the systems, the optimum conditions, including the concentration of PEG and $(NH_4)_2SO_4$, the flotation time and the flow rate of nitrogen, were obtained.

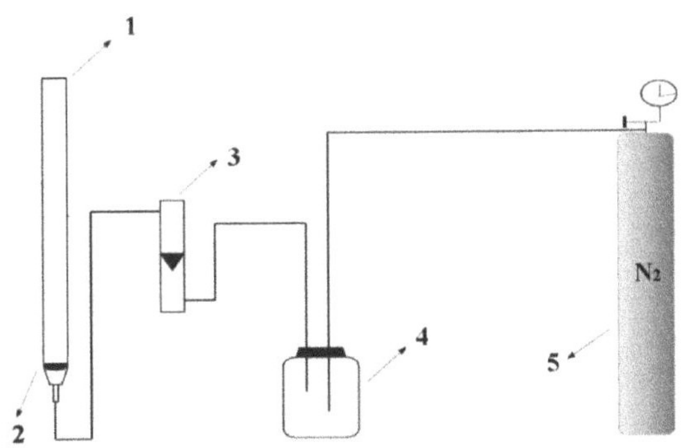

Figure 1. The flotation unit consisted of 1. flotation column; 2. G4 glass core; 3. flowrator; 4. gas buffer device reformed by float type oxygen inhaler; 5. high purity nitrogen cylinder.

2.4. Experimental Design

The response surface method (RSM) experiment was applied to optimize the interaction between the parallel factors and separation parameters. The effects of various factors, including X_1 (the mass fraction of PEG), X_2 (the mass fraction of $(NH_4)_2SO_4$), X_3 (the flow rate of nitrogen) and X_4 (the flotation time), were studied. The experimental design was shown in Table 1. Y and P were taken as the responses and each experiment was replicated three times

Table 1. The factors and levels in the response surface design for optimizing the separation of ovalbumin (OVA) by aqueous two-phase flotation (ATPF).

Variables	Coded Variable Levels		
	−1	0	1
X_1 PEG1000 (*w/w*)%	70	80	90
$X_2(NH_4)_2SO_4$ (*w/w*)%	26	28	30
X_3 flow rate of nitrogen (mL/min)	25	30	35
X_4 flotation time (min)	20	30	40

The top phase of ATPF and salted egg white solution were dialyzed against deionized water overnight by a dialysis bag with 13,000 Da molecular weight cutoffs to remove the PEG 1000 and salt. Then, the dialysate was lyophilized to obtain the purified OVA.

2.5. Determining of Protein

The Bradford [23] was used to determine the concentration of total protein in the top phase. RP-HPLC was used to determine the concentration of OVA in salted egg white solution and in the top phase. The waters C_8 column (4.6 mm × 150 mm) was used and mobile phase A was 0.05% of aqueous TFA, while the mobile phase B was 0.05% TFA in acetonitrile. The flow rate was set to 1.00 mL/min while the concentration of mobile phase A decreased from 93% to 30% in 17 min and increased to 93% before 22 min [24]. The effluent was detected at a wavelength of 280 nm.

2.6. Definition of the Distribution of Protein in ATPF

The flotation efficiency (*Y*) and purification (*P*) of OVA were used to evaluation the enrichment and separation of OVA by ATPF, and were calculated by the following equations:

$$Y = \frac{C_{OVA} \times V_{Top}}{C_{Sample} \times V_{Sample}} \times 100\% \quad (1)$$

$$P = \frac{C_{OVA} \times V_{Top}}{C_{Top} \times V_{Top}} \times 100\% \quad (2)$$

where C_{OVA}, C_{Sample} and C_{Top} were the concentrations of OVA in the top phase and salted egg white solution and total protein in top phase, respectively; while V_{Top} and V_{Sample} were the volume of the top phase and salted egg white solution, respectively.

2.7. Characterization of Ovalbumin Structure

2.7.1. Electrophoresis

The proteins were identified by sodium dodecyl sulfate polyacrylamide gel electrophoresis (SDS–PAGE) in a gel of 12% bis-acrylamide homogeneous. The constant voltage was set at 80 V for approximately 20 min for stacking the gel and at 120 V for 55 min to separate the gel. At the end of the electrophoresis, the tape was stained with Coomassie Brilliant Blue R-250 for 30 min and decolorized with eluent [25].

2.7.2. Nano LC-ESI-MS/MS

The digested protein sample was carried out by an HPLC system with a C18 column (75 μm × 80 mm). The HPLC solvent A was 9.5% water, 90% acetonitrile, and 0.5% formic acid. The HPLC solvent B was 97.5%water, 2% acetonitrile, and 0.5% formic acid. The gradient elution process was described as follows. The concentration of mobile phase A was improved from 2% to 90% within 60 min. The column flow rate was approximately 800 nanoliter per minute after splitting.

The typical sample injection volume was 3 μL. The HPLC system was coupled with a linear ion trap mass spectrometer (LTQ, Thermofisher, Shaihai, China), therefore, a sample eluted from the HPLC column was directly ionized by an electrospray ionization (ESI) process and entered into the mass spectrometer. The ionization voltage was often optimized in the instrument tuning process and it ranged from 1.5 kv to 1.8 kv. The capillary temperature was set at 100 °C. The mass spectrometer was set at the data-dependent mode to acquire MS/MS data via a low energy collision induced dissociation (CID) process. The default charge state was 3 and the default collision energy was 33%. One full scan with one microscan with a mass range of 350 amu to 1650 amu was acquired, followed by nine MS/MS scans of the nine most intense ions with a full mass range and three microscans. The dynamic exclusion feature was set as follows: Repeat the count of 1 and the exclusion duration of 1 min. The exclusion width was 4Da [26].

2.7.3. Spectrum Analysis

The OVA structure was analyzed using a fluorescence spectrometer PerkinElmer LS55. After being dissolved in 10 mmol/L phosphate buffer (pH = 7.2) and diluted to the appropriate concentration. The sample was measured at fluorescence excitation wavelengths of 295 nm and 280 nm, respectively. The emission spectrum was scanned from 315 nm to 415 nm. The fluorescence spectroscopy was performed at both slit and excitation slit widths of 5 nm. The ultraviolet spectrophotometer UV-2550 was used to scan the sample solution in the range of 250–360 nm.

Then, 2 mg of the sample powder was mixed with 200 mg of potassium bromide powder, which was ground under an infrared baking lamp for 10 min and pressed in a mold. The FT IR spectra were measured at 400 cm^{-1}~4000 cm^{-1}. The experiment was repeated three times. After the infrared spectra was obtained, the protein conformation was analyzed by Peak Fit v4.12 software [27].

2.8. *Determination of Functional Properties of Ovalbumin*

2.8.1. Oil Binding Capacity (OBC)

Further, 0.100 g of ovalbumin sample and 1 mL of edible oil were mixed in a 10 mL centrifuge tube by shaking, and the mixture was centrifuged at 716× *g* for 30 min, which was weighed as W_1. The top layer of oil was removed and the residual mixture and the tube weighed as W_2. The quality of the sample was recorded as W_{Sample}. The formula for calculating was as follows:

$$\text{OBC} = \frac{W_1 - W_2}{W_{\text{sample}}} \times 100\% \tag{3}$$

2.8.2. Viscosity

The sample solution with a concentration of 10 g/L was prepared in a 30 °C water bath, which was determined by a digital viscometer.

2.8.3. Emulsibility

The measurement was carried out according to the method of Xiong [28]. The sample was dissolved in a 10 mmol/L phosphate buffer solution and the final concentration of the sample was 5 mg/mL. The protein solution was mixed with soybean oil by a volume ratio 9:1 (oil:protein solution) and placed in a high-speed homogenizer at 10,000 rpm for 1 min. Further, 50 μL of the above sample was diluted 100 times with 0.1% (*w*/*v*) SDS, and then measured by an ultraviolet-visible spectrophotometer at 500 nm. The formula for calculating the emulsifying index (*EAI*, m^2/g) was as follows:

$$EAI = \frac{4.303 \times A \times DF}{1000 \times C \times \varphi \times \theta} \tag{4}$$

where *C* was the concentration of protein, *DF* was the dilution factor (100), θ was the proportion of the oil phase in the emulsion (1:10), φ was the cuvette optical path (1 cm) and *A* was the absorbance value.

2.8.4. Foam Capacity

Furthermore, 100 mL of 1% (*w*/*v*) OVA solution was added into a 250 mL beaker, which was dispersed by a high-speed disperser at 10,000 r/min for 2 min. Then, the volume of foam was measured. The foaming capacity (*FC*) was calculated as follows [29]:

$$FC = \frac{(V_0 - 100)}{100} \times 100\% \quad (5)$$

where V_0 was the total volume of liquid and foam at the time of dispersion stop (mL); 100 was the volume of the stock solution (mL).

3. Results and Discussion

3.1. Single-Factor Variable Analysis

The effect of the mass fraction of PEG 1000 and $(NH_4)_2SO_4$, the flow rate of nitrogen and the flotation time were chosen for this study. Based on *Y* and *P*, the optimal conditions of the ATPF were chosen for further investigation. The results were shown in Figure 2.

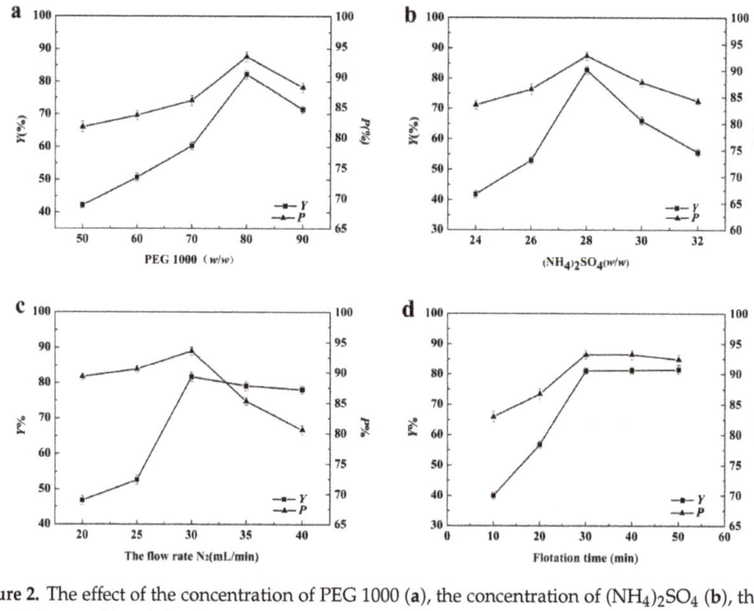

Figure 2. The effect of the concentration of PEG 1000 (**a**), the concentration of $(NH_4)_2SO_4$ (**b**), the flow rate of nitrogen (**c**) and the flotation time (**d**) on the flotation.

The polymer concentration was considered to be an important factor in ATPF. The effect of the mass fraction of PEG 1000 was described in Figure 2a. The trend of *Y* and *P* changed in a similar way. When the PEG 1000 mass fraction ranged from 50% to 80%, *Y* and *P* increased. Generally, electrostatic interactions, hydrophobic interactions and the salting-out effect were the main driving forces of the protein distribution in the ATPS. The hydrophobicity of the top phase increased with the increase of the mass fraction of PEG. OVA was a hydrophobic protein that preferentially interacted with a more hydrophobic PEG-rich phase, which led to a situation that more OVA was separated into the top phase. The high viscosity was believed to be a disadvantage for the bubble entering the top phase through the interface. The viscosity of the polymer solution was affected by the polymer concentration [30], hence, the mass transfer efficiency of the biomolecule was affected. When the mass fraction of PEG was higher than 80%, the viscosity and the bubble duration in the top phase increased to an inappropriate degree and the OVA affinity to the polymer-rich top phase decreased, which caused Y and P to decrease significantly. In addition, the increasing concentration of PEG enhanced the excluded volume of the polymers, which in turn, reduced the solvent volume fraction. Thus, OVA was excluded by PEG in the top phase.

In ATPF, the ratio of the salt phase volume to the polymer phase volume was usually very large, thus high concentrations of salt was necessary for the system to sustain a stable immiscible two-phase [31]. It could be observed that the Y and P increased with the increase of the concentration of $(NH_4)_2SO_4$ in Figure 2b. It could be attributed to the salting-out effect. The solubility of protein in the bottom phase decreased with the increase of $(NH_4)_2SO_4$ concentration, which resulted in the transfer of protein to the top phase. When $(NH_4)_2SO_4$ was above 28%, protein was precipitated at the phase interface, which resulted in the decrease in the Y and P [32].

The gas flow velocity significantly affects the flotation efficiency [33]. Figure 2c showed the effect of the nitrogen flow rate (20–40 mL/min) on the Y and P of OVA. Generally, the low flow rate led to the low mass transfer efficiency. The Y and P of OVA increased when the flow rate ranged from 20 mL/min to 30 mL/min, indicating that increasing of the flow rate of nitrogen made more OVA enter the top phase. At the flow rate of 30 mL/min, the maximum Y and P reached 81.66% and 93.58%, respectively. When the flow rate was above 30 mL/min, the Y and P decreased. The interface between two phases was destroyed by an excessively high flow rate and the two phases were mixed. In addition, the excessively high flow rate resulted in excessive rising speed of bubbles, causing unstable adsorption of the molecules with the bubbles and short contact time [34]. Therefore, the nitrogen flow rate of 30 mL/min was the optimum flotation condition.

The effect of flotation time on the Y and P of OVA in the range of 10–50 min was investigated and the result was described in Figure 2d. The Y and P increased with the flotation time ranging from 10 to 30 min. Before the mass transfer process was at equilibrium, the increasing flotation time facilitated full contact between the bubbles and OVA, which caused more OVA to be transferred to the top phase while the Y and P increased. When the flotation time was 30 min, the maximum Y and P were obtained, and the system basically reached the balance. There was no significant change in Y or P with the increasing of the flotation time after 30 min. Therefore, 30 min was selected as the optimum flotation condition.

From the above analysis, it could be concluded that the best single-factor conditions were as follows: PEG 80% (*w*/*w*) 5 mL, ammonium sulfate 28% (*w*/*w*) 28 mL, the flow rate of nitrogen 30 mL/min and the flotation time 30 min.

3.2. Response Surface Analysis

3.2.1. Statistical Analysis and Model Fitting

The significant independent factors or their interactions of a multivariate complex system (e.g., ATPS) need to be analyzed by an effective mathematical statistical method [35,36]. The response surface methodology (RSM) was applied for the optimization of significant factors.

The experimental results of the response surface design were described in Table 2. The regression analysis was performed on the experimental data obtained using Design-Expert 8.0.6 software, and the equations for predicting the Y and P of OVA were obtained, which were given as follows:

$$Y = 81.81 + 4.66 \times A + 1.18 \times B + 4.48 \times C + 8.14 \times D + 1.99 \times A \times B + 3.86 \times A \times C + 3.44 \times A \times D + 1.39 \times B \times C + 2.81 \times B \times D - 1.42 \times C \times D - 14.38 \times A^2 - 13.69 \times B^2 - 14.83 \times C^2 - 8.66 \times D^2$$

$$P = 93.23 + 1.16 \times A - 1.37 \times B - 2.06 \times C + 0.94 \times D + 1.26 \times A \times B - 0.075 \times A \times C - 0.16 \times A \times D + 1.18 \times B \times C + 0.81 \times B \times D - 0.94 \times C \times D - 2.41 \times A^2 - 2.65 \times B^2 - 4.93 \times C^2 - 1.16 \times D^2$$

Table 2. Box-Behnken (BBD) and the results (means of triplicate tests) for the recovery yield and purification factor of ovalbumin (OVA).

Number	A:PEG 1000 (w/w)%	B:$(NH_4)_2SO_4$ (w/w)%	C:Flow Velocity (mL/min)	D:Flotatiom Time (min)	Y (%)	P (%)
1	1	0	0	-1	50.82	89.82
2	−1	1	0	0	47.44	84.12
3	0	0	0	0	82.56	93.03
4	0	0	0	0	81.15	93.06
5	0	−1	0	−1	53.43	90.61
6	0	0	0	0	81.97	93.08
7	−1	0	−1	0	48.11	86.53
8	0	1	0	−1	50.38	86.23
9	−1	0	1	0	49.26	83.13
10	0	1	1	0	60.27	83.56
11	1	1	0	0	61.97	88.88
12	0	1	−1	0	47.67	85.19
13	0	1	0	1	72.32	90.28
14	1	0	0	1	74.31	91.52
15	−1	−1	0	0	49.09	89.73
16	1	−1	0	0	55.64	89.45
17	−1	0	0	1	59.04	89.64
18	0	−1	1	0	55.31	83.56
19	0	0	−1	1	63.42	90.98
20	0	−1	−1	0	48.29	89.91
21	0	0	0	0	81.78	93.09
22	1	0	1	0	66.01	85.52
23	0	0	0	0	81.58	93.88
24	0	−1	0	1	64.13	91.41
25	1	0	−1	0	49.43	89.22
26	0	0	−1	−1	44.64	87.89
27	0	0	1	−1	55.67	84.93
28	0	0	1	1	68.76	84.27
29	−1	0	0	−1	49.31	87.28

3.2.2. Analysis of Variance

The analysis of variance (ANOVA) for the Y and P models was shown in Tables 3 and 4, respectively. The regression models were highly significant ($p_1 < 0.01$, $p_2 < 0.01$), while the lack-of-fit tests were not significant ($p_1 = 0.0725 > 0.05$, $p_2 = 0.3296 > 0.05$). The determination coefficients (${R_1}^2$ and ${R_2}^2$) of the predicted models were 0.9918 and 0.9558, indicating a high degree of correlation between the true and predicted values. Thus, the models explained the response adequately and could be used to analyze and predict the optimal separation conditions of OVA.

Table 3. The variance analysis of the fitted quadratic polynomial prediction model of Y.

Source	Sum of Squares	df	Mean Square	F	p_1
Model	4430.17	14	316.44	457.18	<0.0001
Residual	9.69	14	0.69	–	–
Lack of fit	8.61	10	0.86	3.20	0.1369
Pure error	1.08	4	0.27	–	–
CV%	–	–	1.38	–	–
${R_1}^2$	–	–	0.9978	–	–

Table 4. The variance analysis of the fitted quadratic polynomial prediction model of p.

Source	Sum of Squares	df	Mean Square	F	p_2
Model	306.56	14	21.90	101.07	<0.0001
Residual	3.03	14	0.22	–	–
Lack of fit	2.50	10	0.25	1.87	0.2856
Pure error	0.53	4	0.13	–	–
CV%	–	–	0.53	–	–
${R_1}^2$	–	–	0.9902	–	–

3.2.3. Interactive Analysis

The relationship between the response and experimental levels of each variable was visualized in a three-dimensional (3D) response surface plot, which provided a method to directly observe the interactions between the two test variables. The factors influencing the *Y* and *P* of OVA were shown as the response surface plots in Figure 3.

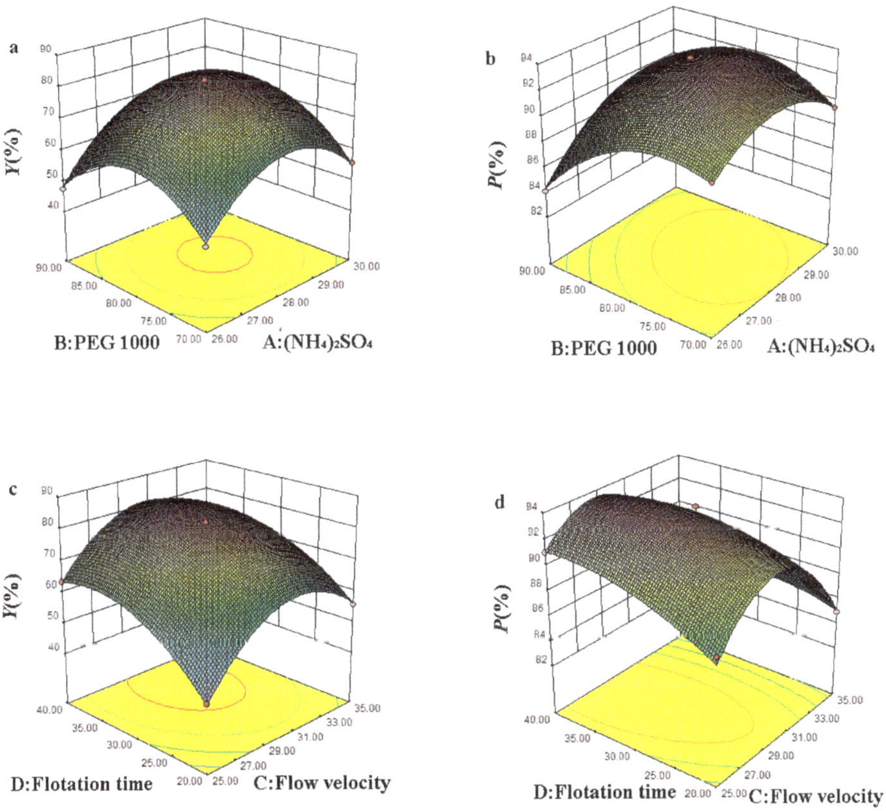

Figure 3. The response surface plots for *Y* (**a**,**c**) and *P* (**b**,**d**) of OVA.

Figure 3a,b showed that *Y* and *P* first increased and then decreased with the increase of the mass fraction of $(NH_4)_2SO_4$, when the mass fraction of PEG was at a specific level. Similarly, *Y* and *P* first increased and then decreased with the increase of the mass fraction of PEG, when the mass fraction of $(NH_4)_2SO_4$ was at a specific level in the ATPF. The *Y* and *P* reached the maximum, when the mass fraction of PEG and $(NH_4)_2SO_4$ reached specific levels. When the mass fraction of PEG and $(NH_4)_2SO_4$ were set at low levels, a weak hydrophobic interaction and a weak ionic strength were obtained in the ATPF, resulting in that less OVA was separated in the top phase. The hydrophobic interaction and ionic strength gradually increased with the increase of the mass fraction of PEG and $(NH_4)_2SO_4$, and hydrophobic OVA was separated into the top phase and other proteins were retained in the bottom phase, meaning that *Y* and *P* were increased. However, the ionic strength in the bottom phase increased and the volume of the top phase decreased with the mass fraction of PEG and the $(NH_4)_2SO_4$ increased, causing the protein to precipitate at the interface of the two phases.

The effect of the interaction between the flow rate and the flotation time on the *Y* and *P* of OVA were shown in Figure 3c,d. It could be inferred that the higher flow rate could reduce the flotation time required to reach equilibrium. However, the interface of the two phases was destroyed when the flow rate was high. In addition, the excessively high flow rate led to excessive rising velocity of the bubbles which caused a short contact time and unstable adsorption of OVA with the bubbles. This might result in the protein adsorbed on the surface of the bubbles falling back into bottom phase. Thus, it was necessary to select the maximum flow rate and the shortest time for ATPF, under the premise of ensuring the separation effect.

3.2.4. Validation of the Best Extraction Conditions

According to the results of Box-Behnken (BBD), the optimal conditions were obtained when the mass fraction of PEG 1000 and $(NH_4)_2SO_4$, the flow rate of nitrogen and the flotation time were 79.22% (*w*/*w*), 28.50% (*w*/*w*), 29.09 mL/min and 34.59 min, respectively. Under these conditions, the Y and P value of OVA could reach 82.73% and 93.89%, respectively. In order to facilitate the operation, the predicted optimal process conditions were amended as follows: The mass fraction of PEG 1000 was 80% (*w*/*w*), the mass fraction of $(NH_4)_2SO_4$ was 28% (*w*/*w*), the flow rate of nitrogen was 30 mL/min and the flotation time was 30 min. Under these conditions, Y was 82.15 ± 0.24% and P was 92.98 ± 0.68%. There was no significant difference between the result and the predicted value, indicating that the model was reliable.

3.3. *Characterization of Ovalbumin Extracted Directly from Salt Egg White*

The result of electrophoresis analysis was shown in Figure 4. The SDS–PAGE of purified OVA from the top phase of ATPF showed one band (lanes B, C) whose molecular mass was approximately 45 kDa. The bands (lanes B, C) were similar to the ovalbumin standard (lanes D). Compared with the SDS–PAGE of OVA purified from the top phase of ATPF, the salted egg white solution showed multiple bands (lane A). According to the electropherogram analysis, it was inferred that OVA could be separated to the top phase by ATPF.

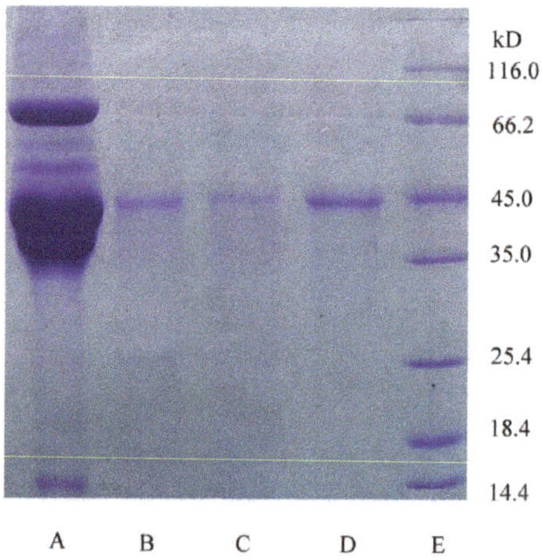

Figure 4. The SDS–PAGE of OVA. Lane A, the salted egg white solution (10 μg); lane B, sample (20 μg) separated by ATPF; lane C, sample (10 μg) separated by ATPF; lane D, OVA standard (10 μg); lane E, molecular mass standards.

The OVA purified from the top phase of the ATPF and the OVA standard were measured by RP-HPLC, and the chromatograms were shown in Figure 5. The retention time of the main peak of the top phase in the chromatogram was approximately 13.5 min, which corresponded to the retention time of the OVA standard. Therefore, the result was consistent with electrophoresis, meaning that OVA was selectively separated to the top phase by the ATPF.

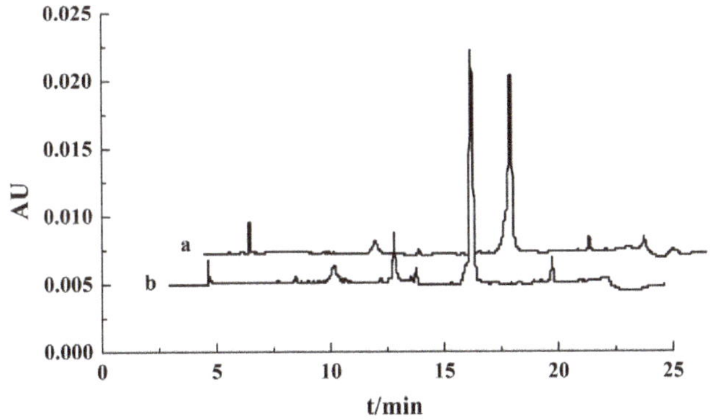

Figure 5. The RP-HPLC chromatograms of the OVA standard (**a**) and OVA purified from the top phase of the ATPF (**b**).

Furthermore, Nano LC-ESI-MS/MS is a reliable and sensitive method for identifying gel-separated protein tapes. Moreover, a single protein tape can be identified in mixed proteins by this method [37]. Tables 5 and 6 showed the results of the OVA purified from the phase of ATPF determined by Nano LC-ESI-MS/MS. There were two proteins in the sample, including OVA and alpha-1-acid glycoprotein. The relative abundance of OVA was 99.4%, meaning that OVA was the main protein in the top phase of ATPF.

Table 5. Nano LC-ESI-MS/MS results of the OVA purified from top phase of ATPF.

Hits	Protein Mass	No. of Peptide	Protein	UniprotKB Databases	Relative Abundance	Probability
1	43,195.66	17	OVA of chick	P01012	99.4%	99.0%
2	22,535.07	3	Alpha-1-acid glycoprotein of chick	Q8JIG5	0.6%	99.0%

Table 6. The peptide of the OVA purified from top phase of ATPF.

Scan No.	Peptide Mass	Peptide Sequence of Protein from Chick	Peptide Probability
6633	1772.89	ISQAVHAAHAEINEAGR	96.2%
6896	887.56	IKVYLPR	87.6%
6955	1554.71	AFKDEDTQAMPFR	96.6%
7045	1580.71	LTEWTSSNVMEER	93.2%
7037	943.53	DILNQITK	89.5%
7155	1354.65	PNDVYSFSLASR	93.9%
7195	1686.83	GGLEPINFQTAADQAR	94.9%
7241	2007.94	EVVGSAEAGVDAASVSEEFR	95.7%
7277	1246.62	ADHPFLFCIK	83.1%
7295	1344.73	HIATNAVLFFGR	95.7%
7490	1521.79	YPILPEYLQCVK	90.7%
7618	2280.17	DILNQITKPNDVYSFSLASR	91.1%
7661	1481.75	PVQMMYQIGLFR	92.3%
7735	2283.14	VTEQESKPVQMMYQIGLFR	91.0%
8705	2459.31	NVLQPSSVDSQTAMVLVNAIVFK	77.0%
10304	1857.96	ELINSWVESQTNGIIR	92.4%
9088	3032.51	VHHANENIFYCPIAIMSALAMVYLGAK	77.6%

The UV spectrum of the OVA standard and OVA purified from the top phase of ATPF were shown in Figure 6. The maximum absorption peak of the ovalbumin obtained from ATPF was approximately

280 nm, which was basically consistent with the OVA standard. The same result was shown in the fluorescence analysis in Figure 7. Thus, it was inferred that there were no significant changes on the spatial structure of the ovalbumin during the separation process.

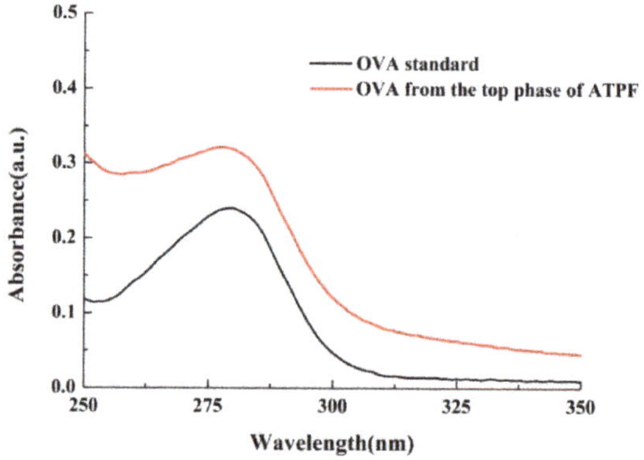

Figure 6. UV spectrum of samples.

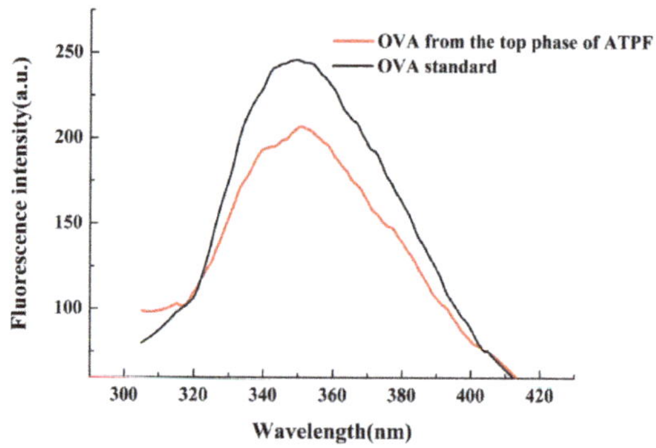

Figure 7. The fluorescence spectrum of the samples.

The secondary structure information of proteins from the top phase of ATPF was analyzed by FT-IR and PeakFit v4.12 software. The protein structure generally corresponds to the amide I absorption band between 1690 cm^{-1} and 1600 cm^{-1} in the infrared spectrum. The corresponding relationship between each sub-peak and secondary structure was as follows: Unordered (1640~1650 cm^{-1}); β-turn (1660~1700cm^{-1}); β-sheet (1610~1640 cm^{-1}); α-helix (1650~1660 cm^{-1}) [38]. Referring to the literature, PeakFit v4.12 software was used to calculate the original infrared map amide I absorption band of the OVA standard and the sample, and the fitted map was obtained after baseline correction, deconvolve Gaussian instrument response functions (IRF) deconvolution and second derivative fitting. According to the relationship between the secondary structure and sub-peak, the percentage of the secondary structure was acquired, as shown in Figure 8. It showed that unordered, β-turn, β-sheet and α-helix were not obviously changed, meaning that there was no change in the spatial structure of OVA in the ATPF separation process.

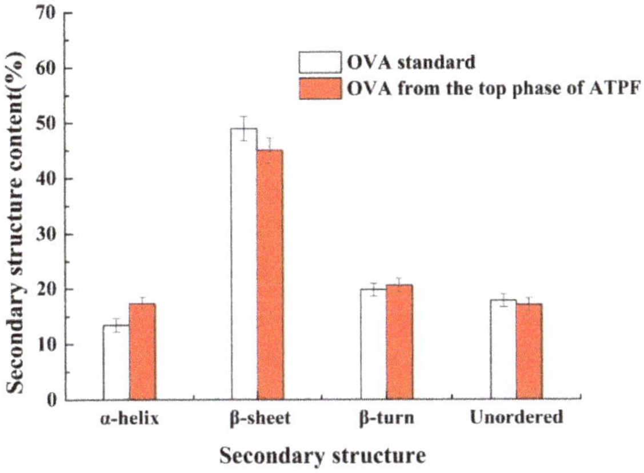

Figure 8. The content of the secondary structure for OVA obtained from different sources.

3.4. Determination of Ovalbumin Functional Properties

The results of functional properties of ovalbumin were shown in Figure 9. There were no significant differences between OVA from the top phase of ATPF and the OVA standard in an oil binding capacity, viscosity, emulsibility and foam capacity. Therefore, the spatial structure of OVA did not change excessively in ATPF and the functional properties of the protein were maintained, meaning that OVA separated from the salted egg white by ATPF could be applied in practice.

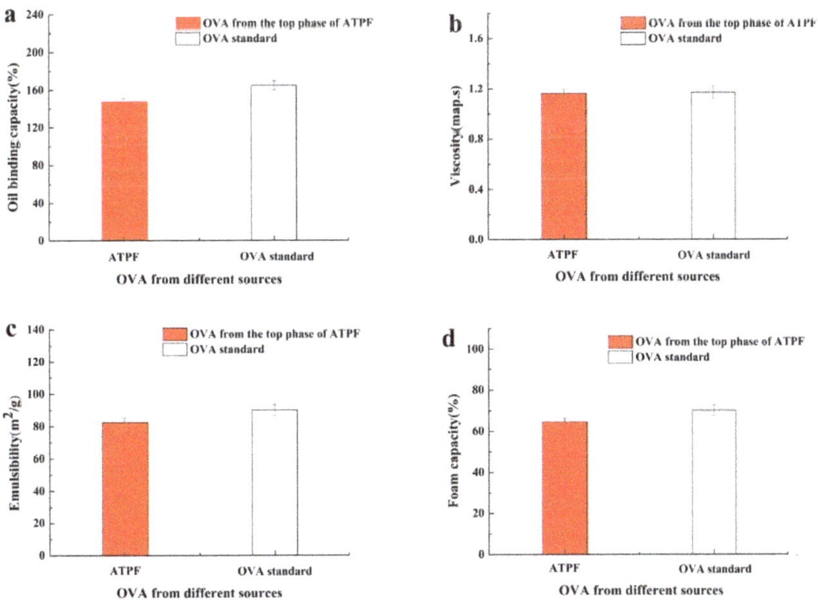

Figure 9. (**a**) Oil binding capacity, (**b**) viscosity, (**c**) emulsibility and (**d**) foam capacity of OVA obtained from different sources.

In this study, the separation and purification of OVA from salted egg whites using an ATPF consisting of PEG 1000/$(NH_4)_2SO_4$ was investigated successfully. The optimum ATPF conditions were obtained using response surface methodology and a Box–Behnken experimental design. Accordingly,

2 mL of the salted egg white solution (salted egg white (*v*):water (*v*) = 1:4) was loaded on the ATPF. Under optimum conditions, including 5 mL of 80% PEG 1000 (*w*/*w*), 28 mL of 28%$(NH_4)_2SO_4$ (*w*/*w*), 30 mL/min of the flow rate and 35 min of the flotation time, the maximum *Y* and the *P* reached 82.15 ± 0.24% and 92.98 ± 0.68%, respectively. The results of SDS–PAGE, RP-HPLC, Nano LC-ESI-MS/MS, ultraviolet spectrum, fluorescence spectrum, FT-IR and functional properties showed that OVA was separated and purified by ATPF, and its functional properties were maintained in ATPF, meaning that ATPF could be a valuable protocol for the separation of OVA from salted egg whites. It might be a sustainable and effective way for the utilization of salted egg whites to reduce environmental pollution.

Author Contributions: Data curation, J.N. and L.W.; formal analysis, L.W.; funding acquisition, B.J. and Z.F.; investigation, D.L.; methodology, B.J., J.N. and C.L.; project administration, B.J. and Z.F.; writing—original draft, J.N.; writing—review & editing, B.J.

Funding: This research was funded by National Key Research and Development Program of China (2018YED0400304) and Natural Science Foundation of Heilongjiang Province (C2018019).

Conflicts of Interest: The authors declare no conflict of interest.

References

1. Jiang, B.; Na, J.X.; Wang, L.L.; Li, D.M.; Liu, C.H.; Feng, Z.B. Eco-innovation in reusing of food by-products: Separation of ovalbumin from salted egg white using aqueous two-phase system of PEG 1000/$(NH_4)_2SO_4$. *Polymers* **2019**, *11*, 238. [CrossRef] [PubMed]
2. Xu, L.; Zhao, Y.; Xu, M.; Yao, Y.; Wu, N.; Du, H.Y.; Tu, Y.G. Changes in physico-chemical properties, microstructure, protein structures and intermolecular force of egg yolk, plasma and granule gels during salting. *Food Chem.* **2019**, *275*, 600–609. [CrossRef] [PubMed]
3. Campbell, L.; Raikos, V.; Euston, S.R. Modification of functional properties of egg-white proteins. *Mol. Nutr. Food Res.* **2003**, *47*, 369–376.
4. Pereira, M.M.; Cruz, R.A.P.; Almeida, M.R.; Lima, A.S.; Coutinho, J.A.P.; Freire, M.G. Single-step purification of ovalbumin from egg white using aqueous biphasic systems. *Process. Biochem.* **2016**, *51*, 781–791. [CrossRef] [PubMed]
5. Benarafa, C.; Remold-O'Donnell, E. The ovalbumin serpins revisited: Perspective from the chicken genome of clade B serpin evolution in vertebrates. *Proc. Natl. Acad. Sci. USA* **2005**, *102*, 11367–11372. [CrossRef] [PubMed]
6. Kovacs-Nolan, J.; Phillips, M.; Mine, Y. Advances in the Value of Eggs and Egg Components for Human Health. *J. Agric. Food Chem.* **2005**, *53*, 8421–8431. [CrossRef] [PubMed]
7. Nisbet, A.D.; Saundry, R.H.; Moir, A.J.G.; Fothergill, L.A.; Fothergill, J.E. The complete amino-acid sequence of hen ovalbumin. *Eur. J. Biochem.* **1981**, *115*, 335–345. [CrossRef]
8. Kratz, F. Albumin as a drug carrier: Design of prodrugs, drug conjugates and nanoparticles. *J. Control. Release* **2018**, *132*, 171–183. [CrossRef] [PubMed]
9. Hopkins, F.G. On the separation of a pure albumin from egg-white. *J. Physiol.* **1900**, *25*, 306–330. [CrossRef]
10. Abeyrathne, E.D.N.S.; Lee, H.Y.; Ahn, D.U. Egg white proteins and their potential use in food processing or as nutraceutical and pharmaceutical agents—A review. *Poult. Sci.* **2013**, *92*, 3292–3299. [CrossRef]
11. Datta, D.; Bhattacharjee, S.; Nath, A. Separation of ovalbumin from chicken egg white using two-stage ultrafiltration technique. *Sep. Purif. Technol.* **2009**, *66*, 353–361. [CrossRef]
12. Geng, F.; Huang, Q.; Wu, X. Co-purification of chicken egg white proteins using polyethylene glycol precipitation and anion-exchange chromatography. *Sep. Purif. Technol.* **2012**, *96*, 75–80. [CrossRef]
13. Zhang, D.D.; Chen, Q.; Hu, L.L.; Chen, X.W.; Wang, J.H. Preparation of a cobalt mono-substituted silicotungstic acid doped with aniline for the selective adsorption of ovalbumin. *J. Mater. Chem. B* **2015**, *3*, 4363–4369. [CrossRef]
14. Liu, J.W.; Wang, M.M.; Zhang, Y.; Han, L.; Chen, X.W.; Wang, J.H. Polymeric ionic liquid modified reduced graphene graphene oxide as adsorbent for highly selective isolation of acidic protein. *RSC Adv.* **2014**, *4*, 61936–61943. [CrossRef]

15. Chen, L.; Zheng, D.H.; Zhang, Y.; Wang, Y.N.; Xu, Z.R. In situ self-assembled reduced graphene oxide aerogel embedded with nickel oxide nanoparticles for the high-efficiency separation of ovalbumin. *J. Sep. Sci.* **2017**, *40*, 1765–1772. [CrossRef]
16. Lee, S.Y.; Khoiroh, I.; Ling, T.C.; Show, P.L. Aqueous two-phase flotation for the recovery of biomolecules. *Sep. Purif. Rev.* **2016**, *45*, 81–92. [CrossRef]
17. Sankaran, R.; Show, P.L.; Yap, Y.J. Sustainable approach in recycling of phase components of large scale aqueous two-phase flotation for lipase recovery. *J. Clean. Prod.* **2018**, *184*, 938–948. [CrossRef]
18. Jiang, B.; Wang, L.L.; Na, J.X.; Zhang, X.Q.; Yuan, Y.Q.; Liu, C.H.; Feng, Z.B. Environmentally-friendly strategy for separation of α-Lactalbumin from whey by aqueous two phase flotation. *Arab. J. Chem.* **2018**. [CrossRef]
19. Pakhale, S.V.; Vetal, M.D.; Rathod, V.K. Separation of bromelain by aqueous two phase flotation. *Sep. Sci. Technol.* **2013**, *48*, 984–989. [CrossRef]
20. Jiang, B.; Na, J.X.; Wang, L.L.; Li, D.M.; Liu, C.H.; Feng, Z.B. Separation and Enrichment of Antioxidant Peptides from Whey Protein Isolate Hydrolysate by Aqueous Two-Phase Extraction and Aqueous Two-Phase Flotation. *Foods* **2019**, *8*, 34. [CrossRef]
21. Bi, P.Y.; Chang, L.; Mu, Y.L. Separation and concentration of baicalin from scutellaria baicalensis georgi extract by aqueous two-phase flotation. *Sep. Purif. Technol.* **2013**, *116*, 454–457. [CrossRef]
22. Chang, L.; Shao, Q.; Xi, X.; Chu, Q.; Wei, Y. Separation of four flavonol glycosides from solanum rostratum dunal using aqueous two-phase flotation followed by preparative high-performance liquid chromatography. *J. Sep. Sci.* **2016**, *40*, 804. [CrossRef] [PubMed]
23. Liu, J.L.; Liu, J.; Zhang, X.F. Production and Characterization of Ovalbumin and Ovotransferrin from Chicken Egg White. *Int. J. Food Eng.* **2012**, *8*, 1–12. [CrossRef]
24. Nau, F.; Mallard, A.; Pages, J.; Brulé, G.; Nau, F. Reversed-phase liquid chromatography of egg white proteins. Optimization of ovalbumin elution. *J. Liq. Chromatogr.* **1999**, *22*, 1129–1147.
25. Jiang, B.; Feng, Z.B.; Liu, C.H.; Xu, Y.C.; Li, D.M.; Ji, G. Extraction and purification of wheat-esterase using aqueous two-phase systems of ionic liquid and salt. *J. Food Sci. Technol.* **2015**, *52*, 2878–2885. [CrossRef] [PubMed]
26. Qin, J.; Zhang, X. Identification of in vivo protein phosphorylation sites with mass spectrometry. *Methods Mol. Biol.* **2002**, *194*, 211–221. [PubMed]
27. Carbonaro, M.; Nucara, A. Secondary structure of food proteins by Fourier transform spectroscopy in the mid-infrared region. *Amino Acids* **2010**, *38*, 679–690. [CrossRef] [PubMed]
28. Xiong, Z.; Zhang, M.; Ma, M. Emulsifying properties of ovalbumin: Improvement and mechanism by phosphorylation in the presence of sodium tripolyphosphate. *Food Hydrocoll.* **2016**, *60*, 29–37. [CrossRef]
29. Li, W.Y.; He, Z.Y.; Xiong, Y.L.; Huang, X.L.; Chen, J. Effects of temperature on the foaming properties of soybean protein isolate. *Sci. Technol. Food Ind.* **2010**, *31*, 86–88.
30. Ling, C. Direct recovery of lipase derived from burkholderia cepacia with aqueous two-phase flotation. *Sep. Purif. Technol.* **2011**, *80*, 577–584.
31. Jiang, B.; Yuan, Y.Q.; Zhang, X.Q.; Feng, Z.B.; Liu, C.H. Separation and Enrichment of Lectin from Zihua Snap-Bean (*Phaseolus vulgaris*) Seeds by PEG 600–Ammonium Sulfate Aqueous Two-Phase System. *Molecules* **2017**, *22*, 1596. [CrossRef] [PubMed]
32. Han, J.; Wang, Y.; Luo, L.; Kang, W.; Chen, H.; Liu, Y.; Li, Y.; Ni, L. Optimization of separation and determination of chloramphenicol in food using aqueous two-phase flotation coupled with HPLC. *J. Iran. Chem. Soc.* **2014**, *11*, 1775–1782. [CrossRef]
33. Lu, Y.; Zhu, X. Solvent Sublation: Theory and Application. *Sep. Purif. Method* **2007**, *30*, 157–189. [CrossRef]
34. Lu, Y.; Chen, B.; Yu, M.; Han, J.; Wang, Y.; Tan, Z.; Yan, Y. Simultaneous separation/enrichment and detection of trace ciprofloxacin and lomefloxacin in food samples using thermosensitive smart polymers aqueous two-phase flotation system combined with HPLC. *Food Chem.* **2016**, *210*, 1–8. [CrossRef] [PubMed]
35. Glyk, A.; Solle, D.; Scheper, T.; Beutel, S. Optimization of PEG salt aqueous two-phase systems by design of experiments. *Chemom. Intell. Lab. Syst.* **2015**, *149*, 12–21. [CrossRef]

36. Rahimpour, F.; Feyzi, F.; Maghsoudi, S.; Hatti, R. Purification of plasmid DNA with polymer-salt aqueous two-phase system: Optimization using response surface methodology. *Biotechnol. Bioeng.* **2006**, *95*, 627–637. [CrossRef]
37. Jiang, B.; Wang, X.J.; Wang, L.L.; Lv, X.M.; Li, D.M.; Liu, C.H.; Feng, Z.B. Two-Step Isolation, Purification, and Characterization of Lectin from Zihua Snap Bean (Phaseolus vulgaris) Seeds. *Polymers* **2019**, *11*, 785. [CrossRef]
38. Qi, B.K.; Ding, J.; Wang, Z.J.; Li, Y.; Ma, C.G.; Chen, F.S.; Sui, X.N.; Jiang, L.Z. Deciphering the characteristics of soybean oleosome-associated protein in maintaining the stability of oleosomes as affected by pH. *Food Res. Int.* **2017**, *100*, 551–557. [CrossRef]

foods

MDPI

Article

Myrtus Communis Liquor Byproduct as a Source of Bioactive Compounds

Fabio Correddu [1,†], Mariateresa Maldini [2,†], Roberta Addis [2], Giacomo Luigi Petretto [2], Michele Palomba [2], Gianni Battacone [1], Giuseppe Pulina [1], Anna Nudda [1,*] and Giorgio Pintore [2]

1 Dipartimento di Agraria, Sezione di Scienze Zootecniche, University of Sassari, viale Italia, 39, 07100 Sassari, Italy

2 Dipartimento di Chimica e Farmacia, University of Sassari, via F. Muroni, 23/b, 07100 Sassari, Italy

* Correspondence: anudda@uniss.it; Tel.: +39-079-229390

† These authors contributed equally to the work.

Received: 27 May 2019; Accepted: 27 June 2019; Published: 30 June 2019

Abstract: The fatty acid (FA), polyphenol content and evaluation of the antioxidant capacity of exhausted *Myrtus communis* berries (EMB) resulting from the production of myrtle liqueur were assessed. All parts of the exhausted berries exhibited high concentrations of carbohydrates, proteins, lipids and phenolic compounds. The lipid fraction contained a high amount of poly unsaturated fatty acids (PUFA), mainly represented by linoleic acid (>70%). Of the phenolic acids evaluated by liquid chromatography/mass spectrometry, ellagic acid was the most predominant (>50%), followed by gallic and quinic acids. Quercetin and quercetin3-O-rhamnoside were the most abundant flavonoids. The seed extracts showed a higher antioxidant potential than the pericarp extracts; the same trend was observed for total phenolic compounds evaluated by spectrophotometric assay. The overall high content of bioactive compounds and the high antioxidant potential of this byproduct sustain its suitability for a number of industrial applications, such as a food ingredient in novel foods, an additive in cosmetic formulations or a component of animal feed formulations.

Keywords: LC-MS/MS; fatty acids; polyphenols; antioxidant activity

1. Introduction

Over recent years, interest in the recovery of high added-value products from waste plant material has grown worldwide, as the re-use and valorization of these byproducts have become important economic issues in a number of industrial sectors [1].

Food processing waste often consists of organic material, the disposal of which presents a serious pollution risk. However, the appropriate management and disposal of such materials entails additional cost. Attempting to extract extra value out of agricultural waste is thus a major step towards alleviating this problem. Many byproducts arising from the processing of fruit and vegetables are rich in phytochemicals that may still retain valuable chemical and biological properties [2]. For example, it has been repeatedly demonstrated that such byproducts can possess high amounts of phenolic compounds—a large group of secondary metabolites that includes: phenolic acids, flavonoids, anthocyanins, and proanthocyanidins [3]. These metabolites have received a great deal of attention because of their numerous biological properties, such as their anti-mutagenic, cardioprotective, anti-inflammatory, anti-carcinogenic, anti-allergic, antiviral and antioxidant activities [4]. Indeed, myrtle liqueur itself has been reported to exhibit strong antioxidant activity [5]. Furthermore, several epidemiological studies suggest that a diet rich in antioxidants may have a positive impact by increasing the reactive antioxidant potential of an organism and reducing the risk of certain degenerative diseases that originate from deleterious free radical reactions [6].

Polyphenol-rich byproducts could be used as functional antioxidant ingredients in the food industry; this possibility is particularly interesting because the currently available and widely used synthetic antioxidants, such as butylated hydroxyanisole (BHA) and butylated hydroxytoluene (BHT), have been suspected to cause negative health effects [7]. Polyphenol-rich byproducts could also provide attractive solutions for the pharmaceutical and cosmetic industries [8,9].

The utilization of byproducts rich in polyphenols as feedstuff has also been explored since recent studies into ruminants showed beneficial effects on health conditions [10], protein and lipid ruminal metabolism [11,12] and the immune system [13], as well as an enhancement of milk production and quality [14,15]. In addition, their use as feed in animal diets may help avoid expensive byproduct treatments, which often lead to further waste production [1].

To the best of our knowledge, no studies have been performed on the byproducts derived from myrtle liqueur production. The berries of *Myrtus communis* are used to produce a sweet myrtle liqueur by their hydroalcoholic infusion (> 40 °C) lasting at least 15 days [16]. More than three million bottles of myrtle liqueur are currently produced in Sardinia per year, and it is fast becoming one of the most popular Sardinian exports [17]. Of consequence, Sardinia produces a considerable amount of exhausted berries of *Myrtus communis* as a waste product, estimated to be approximately 200,000 tons/year.

The present study aimed to characterize the chemical composition and the phenolic profile of exhausted berries of *Myrtus communis* resulting from myrtle liqueur production and to test their antioxidant activity in order to evaluate their potential for further exploitation.

2. Materials and Methods

2.1. Reagents and Standards

The solvents used for extraction (methanol, acetonitrile and formic acid) were purchased from Sigma-Aldrich Chemical Company (St Louis, MO, USA). A Milli-Q purification system (Millipore, Bedford, MA, USA) was used to prepare high-performance liquid chromatography (HPLC) grade water. Standards of gallic acid, caffeic acid, p-coumaric acid, ellagic acid, ferulic acid, sinapic acid, quinic acid, syringic acid, chlorogenic acid, phloridzin, kaempferol, luteolin, quercetin, isorhamnetin, myricetin, apigenin, epicatechin and catechin were purchased from Sigma-Aldrich. Standards of quercetin rhamnoside, isoquercetin, rutin, robinin, isorhamnetin rutinoside, neohesperidin, quercetin galactoside, myricitrin, myricetin galactoside, epigallocatechin, epigallocatechin gallate, procyanidin B1, procyanidin B2, cyanidin-3-O-glucoside, cyanidin-3-O-arabinoside, cyanin, delphinidin-3-O-glucoside, malvidin-3-O-glucoside, and pelargonidin-3-O-glucoside chloride were purchased from Extrasynthese (Genay, France).

The reference standard mixture of 37 FAME (FAME mix 37) was acquired from Supelco (Sigma-Aldrich, Bellefonte, PA, USA); other reference standards were purchased from Matreya Inc. (Pleasant Gap, PA, USA): PUFA-2, a nonconjugated 18:2 isomer mixture of individual PUFA, eicosapentaenoic acid, (EPA), docosahexaenoic acid (DHA), arachidonic acid (ARA), C18:3 cis-6,9,12, and C18:3 cis-9,12,15.

2.2. Samples Collection and Extracts Preparation

The exhausted myrtle berries (EMB) analyzed were obtained from a local distillery. The EMB were dried at ambient temperature for a week, successively, in an air oven at 40 °C until complete drying (24 h) and stored at 0–5 °C for later uses. For analysis, seeds were manually separated from pericarps by screening after air-drying, and samples of whole EMB, seeds and pericarps were finely ground.

For polyphenolic analysis and antioxidant assays, the following extraction procedure was employed: the samples were sonicated for 60 min in a solution of 70:30 ethanol:water (*v/v*) with a sample to solvent ratio of 13:25 (*w/v*) and kept in the dark overnight. After filtration, a rotary evaporator was used to remove completely the solvent. Ultrapure water at the same volume of extraction was used to dissolve the dried samples that were then filtered using 0.20-μm syringe PVDF filters (Whatmann International Ltd., Maidstone, UK).

2.3. Chemical Composition

The dry matter (DM) content of the samples was determined by oven-drying at 105 °C for 24 h. The fiber fractions content (neutral detergent fiber, NDF; acid detergent fiber, ADF; and acid detergent lignin, ADL) was determined following the sequential procedure described by Van Soest, Robertson and Lewis [18], using the filter bag equipment of Ankom (Ankom Technology Corp., Fairport, NY, USA). Ash, protein (CP) and ether extract (EE) contents were determined following the analytical procedures (methods 942.05, 988.05 and 920.39, respectively) reported by AOAC [19,20]. Organic matter (% DM) was calculated as follow: 100 – ash. NFC (non-fiber carbohydrate) was calculated as follows: NFC (% DM) = 100 − (NDF + CP + ash + EE). Hemicelluloses and cellulose were calculated as NDF – ADF and ADF – ADL, respectively. Carbohydrates and gross energy were calculated according to Guimarães, Barros, Carvalho and Ferreira [21] as follow: carbohydrates (% DM) = 100 − (CP + ash + EE); and energy (kcal/100 g DM) = 4 × (CP + carbohydrate) + 9 × (EE). These parameters (except for energy) were expressed as percentage of DM. Analyses were carried out in triplicate, and results were reported as mean ± SD.

2.4. Fatty Acid Profile

The FA profiles of seeds, pericarps and whole EMB were determined following the method of Kramer, Fellner, Dugan, Sauer, Mossoba and Yurawecz [22] with some modifications. The powder was processed with 2 mL of 0.5 M methanolic sodium methoxide (Sigma-Aldrich, Spain) kept in a water bath at 50 °C for 10 min and then cooled at room temperature. The samples were then processed (in a water/ice bath) with 3 mL of HCl/methanol (3M) prepared, freshly, with acetyl chloride and methanol. The samples were heated again in a water bath at 50 °C for 10 min and cooled to room temperature. After adding 1 mL of solution containing methyl nonadecanoate (C19:0) as internal standard (Sigma Chemical Co., St. Louis, MO, USA) and 7.5 mL of K_2CO_3 (0.43 M) the samples were shaken and centrifuged (1500× *g*, room temperature, 5 min), and supernatant was kept in an amber vial for GC analysis. The Fatty acid methyl esters (FAME) determination was carried out using a 7890A GC System (Agilent Technologies, Santa Clara, CA, USA), provided with an autosampler (7693, Agilent Technologies, Santa Clara, CA, USA), a split/splitless injection port (split mode, 1:80), and a flame ionization detector (FID). FAME separation was carried out on a capillary column (CP-Sil 88, 100 m × 0.250 μm i.d., 0.25 μm film thickness, Agilent Technologies, Santa Clara, CA, USA). The oven temperature was maintained at 45 °C for 4 min, increased by 13 °C/min to 175 °C, and held for 27 min; finally, it was increased by 4 °C/min until 215 °C, and held for 35 min. Carrier gas (Helium) was used at a flow rate of 1 mL/min and with a pressure of 28 psi. Sample volume injection was 1 μL. Both injector and detector temperatures were 250 °C. Peak detection was operated using OpenLAB CDS GC ChemStation Upgrade software data system (Revision C.01.04, Agilent Technologies Inc., Santa Clara, CA, USA). Identification of individual FAME was carried out by the comparison of their retention times with those of standards methyl ester, and isomeric profiles found in the literature [23]. Analysis were carried out in triplicate, and results were expressed as mean ± SD.

2.5. Antioxidant Capacity

The antioxidant capacity in seeds and pericarp extracts was evaluated by two colorimetric assays measuring the activity of the samples to scavenge the two radicals DPPH (2,2-diphenyl-1-picrylhydrazyl) and ABTS (2,2′-azinobis-(3-ethylbenzothiazoline-6-sulfonic acid).

2.6. DPPH Radical Scavenging Activity

The DPPH radical scavenging assay was performed according to the method proposed by Brand-Williams, Cuvelier and Berset [24] with some modifications as previously reported by

Maldini et al. [25]. The DPPH (2.4 mg) was dissolved in 10 mL of ethanol 70% and stored, in the dark, at −20 °C. An aliquot of 1 mL of this solution was added to 45 mL ethanol 70%, to prepare a work solution having an absorbance of 1.2 ± 0.02, at λ of 517 nm. The ethanolic extracts (100 μL) at different concentrations (from 0.1 to 100 μg) were added to the work solution, to reach 1 mL of final volume. The solutions were then mixed thoroughly and kept in the dark at 25 °C. The extent of DPPH radical reduction was measured by reading the solution's absorbance at 517 nm at zero and after 30 min. A Trolox calibration curve in the range 0.25–7.5 μg/mL was used as positive reference. The spectrophotometer used for the assay was an Ultrospec 4300 pro UV–vis (Amersham Biosciences, Piscataway, NJ, USA), equipped with a temperature controller. Solutions were read in 1 cm quartz cuvette.

The following equation was used to calculate the scavenging activity of the DPPH radical:

$$\% \text{ scavenging of DPPH radical} = [(Ab - As)/Ab] \times 100 \quad (1)$$

where Ab is the absorbance of the control reaction (blank); and As is the absorbance of the hydroalcoholic extracts (sample). Analyses were carried out in triplicate, and results were expressed as mean ± SD.

2.7. ABTS Radical Scavenging Assay

The ABTS radical scavenging assay was performed following the method detailed by Petretto et al. [26]. The assay is based on the properties of an antioxidant compound to reduce the radical cation $ABTS^{\cdot+}$ (chromophore blue/green) to ABTS. The extent of the reduction and the timescale depend on the concentration and the antioxidant power of the considered compound and on the duration of the reaction. The first step was the production of the radical cation obtained by the reaction between ABTS and potassium persulfate (2.45 mM) to reach a final concentration of 7 mM. The solution was kept in the dark at 25 °C for 12–16 h. Before use, the $ABTS^{\cdot+}$ solution was diluted with ethanol 70% to obtain a work solution with an absorbance of 0.7 ± 0.02 at λ of 734 nm. The ethanolic extracts (100 μL at different concentrations (from 0.1 to 100 μg) were added to the work solution, to reach 1 mL of final volume. The reduction of $ABTS^{\cdot+}$ radical cation was recorded (each concentration in triplicate) at zero and after 50 min. Antioxidant capacity of each sample was reported as percent of inhibition. In addition, the IC_{50} value (reported as mean ± SD) was calculated from regression analysis.

2.8. Determination of Total Phenols

Total phenols were measured by a colorimetric assay based on procedures described by Lizcano, Bakkali, Ruiz-Larrea and Ruiz-Sanz [27] with some modifications as previously described [25]. Results were expressed as μg of gallic acid equivalent (GAE) per mg of each EMB part.

2.9. ESI-MS and ESI-MS/MS Analysis

MS analysis was carried out using an ABSciex (Foster City, CA, USA) API4000 Q-Trap spectrometer. Depending on the investigated compound, the spectrometer was set in the both negative/positive ion mode. To optimize the experimental conditions, a solution of each standard (1 μg/mL in methanol:water 50:50) was infused at 10 μL/min into the source.

2.10. HPLC–ESI-MS and HPLC–ESI-MS/MS Analysis

An UHPLC system was used to perform quantitative on-line UHPLC-ESI-MS/MS analyses; the system was interfaced to an ABSciex (Foster City, CA, USA) API4000 Q-Trap instrument in Multiple Reaction Monitoring (MRM) mode, with the mass spectrometer operating as a triple quadrupole analyzer.

Liquid chromatography analysis was conducted using a Flexar UHPLC AS system (Perkin-Elmer, USA). The system was equipped with: autosampler, degasser, pump (Flexar FX-10), and column oven (PE 200). Injection volume of each sample was 5 μL and polyphenolic compounds were separated on a

X Select CSH C18 column (Waters, Milford, MA) (100 mm × 2.1 mm i.d., 2.5 µm d). The temperature was kept at 47 °C and 2 mobile phases were used: A (formic acid 0.1% in H_2O) and B (formic acid 0.1% in acetonitrile). For anthocyanin compounds, a XSelect HSS T3 column (Waters, Milford, MA , USA) (100 mm × 2.1 mm i.d., 2.5 µm d) was selected and elution was carried out at 41 °C. The flow and the solvent gradient used for elution of phenolic compounds and anthocyanins were different and were previously reported in the work of Maldini et al. [28].

For each compound, selected transitions and the optimized parameters were listed in supplementary material (Table S1). Analyst software 1.6.2. was used for the data acquisition and processing.

2.11. Calibration and Quantification of Phenolic Compounds

A stock solution for each standard was prepared at 1mg/mL in methanol:water (50:50). To calculate the calibration curves for each compound, five work solutions at the concentrations of 0.01, 0.05, 0.1, 1, 5 and 10 µg/mL of standards were prepared by diluting the stock solution with methanol. (each work solution was analyzed in 3 replicates).

2.12. Method Validation

Validation of LC–MS/MS method was performed following the guidelines of the European Medicines Agency (EMEA), concerning the analytical methods validation [29].

The determination of the limit of detection (LOD) and the limit of quantification (LOQ) for each standard compound were by the serial dilution of a stock solution until the signal:noise (S/N) ratios were 3:1 and 10:1, respectively. The LOD and LOQ values for each compound are reported in the Supplementary Materials (Table S2).

To evaluate the precision of the method, variations of intraday and interday analysis were assessed as follows: for each sample, 3 aliquots within the same day, and other 3 aliquots during three consecutive days (one per day) were analyzed. The precision of the method was expressed as percentage relative standard deviation (RSD) (Supplementary Materials, Table S2).

The efficiency of extraction and the analytical method were evaluated by performing recovery tests (in triplicate, with the optimized parameters). LC–MS/MS analysis were carried out on samples after the addition of standard solutions (at different concentration). The recovery (%) ranged from 94.6% to 106.7% within the same day.

2.13. Statistical Analysis

The one-way analysis of variance (ANOVA) was used to determine significance differences between seeds, pericarps and whole-EMB. Means were separated using Tukey's test ($p < 0.05$). Differences in total phenols contents over concentrations between seeds and pericarps were assessed using linear regression in which the slope variations were compared with a global test of coincidence using an online statistical calculator (http://www.danielsoper.com/statcalc3/calc.aspx?id=103 [30]). When the data were normally distributed, the association between variables was evaluated by the Pearson product moment correlation coefficient.

3. Results and Discussions

The average proportions of seeds and pericarps in the whole EMB after drying at 40 °C were 58.60 ± 4.8 and 41.40 ± 4.8 (mean ± SD), respectively.

3.1. Chemical Composition

The results of the chemical analyses performed on whole EMB and the separated seeds and pericarps are reported in Table 1. The seeds presented slightly higher values for DM and organic matter than for pericarps ($p < 0.01$), whereas the ash content was higher for the pericarps than for seeds (p

< 0.01). Regarding the fiber content, NDF was higher in seeds than in pericarps ($p < 0.01$), whereas ADF and lignin contents were both higher in pericarps than in seeds ($p < 0.01$). The differences in fiber content and composition for the two EMB fractions was further highlighted by the higher hemicellulose (27.75 vs. 21.24) and cellulose (25.21 vs. 8.62) contents in seeds than in pericarps ($p < 0.01$). Non-fiber carbohydrates (NFC) were more abundant in pericarps than in seeds ($p < 0.01$). The crude protein and fat contents were higher in seeds than in pericarps ($p < 0.01$); in particular, the values for crude protein and fat were about 2-fold and almost 10-fold higher in seeds compared with pericarps, respectively. These differences result in a significantly higher energy value for seeds compared with pericarps (445 vs. 384 kcal/100 g DM, $p < 0.01$). Overall, the chemical composition of whole EMB showed interesting value from a nutritional point of view, suggesting a possible use as feedstuff. This is evidenced by the value of gross energy (425 kcal/100 g of DM), which is comparable to that of typical feeds used in ruminant nutrition, as soybean meal (350–450 kcal/100 g of DM). Recently, the EMB was used as supplement in two nutritional trials in sheep [31,32], evidencing contrasting results in term of milk production (no effect or reduction of milk yield) and milk composition (no effect or reduction of protein and fat content, and reduction or no effect of milk urea content), but both studies agreed on the suitability of this by product as feed in sheep.

Table 1. Chemical composition of seeds, pericarps and whole exhausted myrtle berries.

Chemical Composition [1]	Pericarps	Seeds	Whole EMB	p-Value
dry matter (DM), %	88.58 ± 0.01 ^{c}	91.37 ± 0.03 ^{a}	89.73 ± 0.04 ^{b}	***
organic matter, % of DM	94.46 ± 0.02 ^{c}	98.05 ± 0.03 ^{a}	96.53 ± 0.02 ^{b}	***
NDF, % of DM	62.66 ± 1.81 ^{b}	69.38 ± 0.17 ^{a}	65.14 ± 1.61 ^{b}	**
ADF, % of DM	59.71 ± 1.16 ^{a}	51.67 ± 0.16 ^{b}	53.34 ± 1.37 ^{b}	***
NFC, % of DM	25.21 ± 1.83 ^{a}	8.62 ± 0.17 ^{c}	14.64 ± 2.05 ^{b}	***
ADL, % of DM	38.47 ± 0.05 ^{a}	23.92 ± 0.45 ^{c}	29.85 ± 0.93 ^{b}	***
hemicelluloses, % of DM	2.95 ± 0.65 ^{c}	17.71 ± 0.00 ^{a}	11.80 ± 0.25 ^{b}	***
cellulose, % of DM	21.24 ± 1.20 ^{c}	27.75 ± 0.29 ^{a}	23.49 ± 0.45 ^{b}	***
carbohydrates, % of DM	87.87 ± 0.03 ^{a}	78.00 ± 0.16 ^{c}	79.78 ± 0.48 ^{b}	***
proteins, % of DM	5.35 ± 0.02 ^{b}	9.43 ± 0.15 ^{a}	9.02 ± 0.50 ^{a}	***
fat, % of DM	1.24 ± 0.00 ^{c}	10.61 ± 0.00 ^{a}	7.73 ± 0.00 ^{b}	***
total fatty acids, % of DM	0.39 ± 0.02 ^{c}	9.17 ± 0.07 ^{a}	5.92 ± 0.05 ^{b}	***
ash, % of DM	5.54 ± 0.02 ^{a}	1.95 ± 0.03 ^{c}	3.47 ± 0.02 ^{b}	***
gross energy, kcal/100 g DM	384.04 ± 0.08 ^{c}	445.23 ± 0.10 ^{a}	424.74 ± 0.07 ^{b}	***

Means in the same row with different superscripts differ ($p < 0.05$). Values are means with standard deviation ($n = 3$). ** $p < 0.01$; *** $p < 0.001$. [1] NDF, neutral detergent fiber; ADF, acid detergent fiber; NFC, non-fiber carbohydrates; ADL, acid detergent lignin.

3.2. Fatty Acid Composition

The fatty acid (FA) profiles of pericarps, seeds and whole EMB are presented in Table 2. The FA profiles for EMB and seeds were very similar due to the low contribution of pericarps to the lipid content of whole EMB (see Table 1). The FA profile of pericarps showed a composition similar to that of seeds, but with a different proportion of each FA. Linoleic acid (C18:2 n-6, LA) was the most abundant FA in seeds and in whole EMB, accounting for 75% and 71% of total FA, respectively. The other most representative FAs in seeds and whole EMB were oleic acid (C18:1 cis-9, OA; 9.25% and 9.41%, respectively), palmitic acid (C16:0, PA; 8.30% and 9.34%, respectively), and stearic acid (C18:0, SA; 3.99% and 4.26%, respectively). In pericarps, the most abundant FA was PA, accounting for about 25%, followed by LA, OA, SA, arachidic acid (C20:0, AA) and LNA (17.31%, 11.69%, 8.12%, 5.20% and 4.24%, respectively). Interestingly, pericarps showed a higher proportion of LNA and saturated and unsaturated long chain FAs when compared with seeds ($p < 0.01$). In general, these results are in line with the FA composition of seeds and pericarps of fresh myrtle berries as reported in previous studies [33,34]. However, a different FA profile was reported by Cakir [35], who found the OA content of seeds and mesocarps to be 64% and 72%, respectively, with LA accounting for only 12.7% and 1.7%, respectively. This discordance could be ascribed to a difference in the maturation stage of the berries. In fact, when the variations in FA composition of myrtle berries were studied at different time points during fruit maturation [36], PA and OA were shown to be the most abundant FA in the first stage of ripening (37.03% and 21.89%, respectively), whereas their proportions decreased progressively throughout all stages of ripening (until 13.58 and 6.49%, respectively). On the other hand, the proportion of LA only accounted for 12.21% at 30 days after flowering and increased progressively to 71.34% in fully ripe fruit (180 days post flowering), thus reaching comparable values to those observed in our study.

Table 2. Fatty acid profile (mean of FAME ± SD) of seeds, pericarps and whole exhausted barriers of myrtle resulted from liquor production.

Fatty Acids [1] (g/100 g of FAME)	Pericarps	Seeds	Whole EMB	*p*-Value
C10:0	1.56 ± 0.44 [a]	0.04 ± 0.00 [b]	0.14 ± 0.03 [b]	***
C12:0	0.32 ± 0.06 [a]	0.01 ± 0.00 [b]	0.03 ± 0.00 [b]	***
C14:0	2.47 ± 0.04 [a]	0.05 ± 0.00 [c]	0.21 ± 0.00 [b]	***
C15:0	0.14 ± 0.02 [a]	0.01 ± 0.00 [b]	0.02 ± 0.00 [b]	***
C16:0 (PA)	24.47 ± 0.31 [a]	8.30 ± 0.04 [c]	9.34 ± 0.02 [b]	***
C16:1 *cis*-9	0.36 ± 0.05 [a]	0.02 ± 0.00 [b]	0.04 ± 0.00 [b]	***
C17:0	0.00 ± 0.00 [c]	0.11 ± 0.00 [a]	0.11 ± 0.00 [b]	***
C18:0 (SA)	8.12 ± 0.11 [a]	3.99 ± 0.04 [c]	4.26 ± 0.03 [b]	***
C18:1 *trans*-5	0.68 ± 0.03 [a]	0.06 ± 0.01 [b]	0.10 ± 0.01 [b]	***
C18:1 *trans*-6 + *trans*-8	1.03 ± 0.12 [a]	0.11 ± 0.01 [b]	0.17 ± 0.02 [b]	***
C18:1 *trans*-11	0.95 ± 0.11 [a]	0.03 ± 0.01 [b]	0.09 ± 0.01 [b]	***
C18:1 *cis*-9 (OA)	11.69 ± 0.12 [a]	9.25 ± 0.06 [b]	9.41 ± 0.05 [b]	***
C18:1 *cis*-11	0.88 ± 0.12 [a]	0.45 ± 0.04 [b]	0.48 ± 0.04 [b]	***
C18:1 *cis*-16	0.00 ± 0.00 [b]	0.07 ± 0.00 [a]	0.07 ± 0.00 [a]	***
C18:2 *trans*-11,*trans*-15	1.68 ± 0.39 [a]	0.00 ± 0.00 [b]	0.11 ± 0.03 [b]	***
C18:2 *n*-6 (LA)	17.31 ± 0.57 [c]	75.09 ± 0.18 [a]	71.38 ± 0.15 [b]	***
C18:3 *n*-3 (LNA)	4.24 ± 0.31 [a]	0.42 ± 0.01 [b]	0.67 ± 0.02 [b]	***
C20:0	5.20 ± 0.09[a]	0.58 ± 0.03 [b]	0.88 ± 0.03 [b]	***
C20:1 *cis*-5	0.02 ± 0.04	0.04 ± 0.01	0.04 ± 0.01	NS
C20:1 *cis*-9	0.50 ± 0.31 [a]	0.02 ± 0.00 [b]	0.05 ± 0.02 [ab]	*
C20:1 *cis*-11	0.76 ± 0.25 [a]	0.22 ± 0.00 [b]	0.25 ± 0.02 [b]	**
C20:2 *n*-6	0.30 ± 0.13 [a]	0.11 ± 0.01 [b]	0.13 ± 0.01 [ab]	*
C20:4 *n*-6	1.32 ± 0.26 [a]	0.03 ± 0.00 [b]	0.11 ± 0.02 [b]	***
C22:0	4.03 ± 0.13 [a]	0.11 ± 0.00 [c]	0.36 ± 0.01 [b]	***
C22:3 *n*-6	0.33 ± 0.05 [a]	0.14 ± 0.00 [b]	0.15 ± 0.00 [b]	***
EPA	0.00 ± 0.00 [b]	0.02 ± 0.01 [a]	0.02 ± 0.01 [a]	*
C24:0	3.77 ± 0.15 [a]	0.10 ± 0.01 [c]	0.34 ± 0.01 [b]	***
C24:1 *cis*-15	0.43 ± 0.00 [a]	0.20 ± 0.00 [c]	0.21 ± 0.00 [b]	***
C25:0	0.80 ± 0.09 [a]	0.02 ± 0.00 [b]	0.07 ± 0.01 [b]	***
C26:0	3.61 ± 0.63[a]	0.02 ± 0.00 [b]	0.25 ± 0.04 [b]	***
Σ SFA	55.04 ± 0.97 [a]	13.41 ± 0.09 [c]	16.09 ± 0.02 [b]	***
Σ MUFA	17.31 ± 0.39 [a]	10.47 ± 0.07 [b]	10.91 ± 0.09 [b]	***
Σ PUFA	25.28 ± 1.13 [c]	76.05 ± 0.14 [a]	72.79 ± 0.12 [b]	***
SFA:PUFA ratio	2.18 ± 0.13 [a]	0.18 ± 0.00 [b]	0.22 ± 0.00 [b]	***
Σ C18 1 *trans*	2.66 ± 0.22 [a]	0.20 ± 0.02 [b]	0.36 ± 0.02 [b]	***
Σ C18:2 (except LA)	1.77 ± 0.55 [a]	0.22 ± 0.05 [b]	0.32 ± 0.03 [b]	**
Σ C21:1	1.28 ± 0.07 [a]	0.28 ± 0.01 [b]	0.35 ± 0.01 [b]	**

Means in the same row with different superscripts differ ($p < 0.05$). NS $p > 0.05$; * $p < 0.05$; ** $p < 0.01$; *** $p < 0.001$.
[1] FAME, fatty acid methyl ester; PA, palmitic acid; SA, stearic acid; OA, oleic acid; LA, linoleic acid; LNA, α-linolenic acid; EPA, eicosapentaenoic acid; SFA, total saturated fatty acids; MUFA, total monounsaturated fatty acids; PUFA, total polyunsaturated fatty acids.

The high amount of PUFA makes the lipid content of EMB potentially useful because of the beneficial biological and nutritional properties of these compounds. Indeed, LA could be included in cosmetic formulations since it exhibits important skin protection properties [37]. Moreover, the inclusion of lipid sources (with a high proportion of PUFA) in ruminant diets represents a useful strategy to increase the proportion of beneficial FA in meat and milk and their derived products [38]. Values of LA that exceed 50% of total FA are typical of plant oils, such as soybean, sunflower and grape seed oils [39]. In particular, the FA profile reported in our study for the myrtle seeds is very close to that of the grape seed byproduct [40], which was found to enhance the concentration of beneficial FA in sheep milk when added to the animals' diet [14].

3.3. Polyphenolic Compounds

A preliminary screening of polyphenol total content was performed using the Folin–Ciocalteau method and data were expressed as μg GAE/mg of dry extract; the results are in line with those reported by Wannes and Marzouk [41] relating to fresh berry parts. As evidenced by the results (Table 3), the total polyphenol content was higher in seed extracts than pericarps ($p < 0.01$). In two recent trials on sheep nutrition, the presence of polyphenols in EMB has been associated to the reduction in blood and milk urea concentration [31] and in ammonia accumulation in rumen [42]. It seems correlated to the ability of polyphenols to bind dietary proteins and to reduce their ruminal degradation. In addition, EMB was found to be effective in reducing the proteolytic bacteria in rumen [42]. These findings also point out that *Myrtus* byproduct could be used to increase feed efficiency in animals, in terms of better protein utilization.

Table 3. Determination of total phenols by the Folin–Ciocalteau method in extracted of seeds and pericarps of EMB. Data are expressed as mean ± SD of 3 independent experiments. Each result showed a positive correlation ($p < 0.001$) with DPPH and ABTS results.

Extract Concentration (μg/μL)	Seeds (μg GAE [1])	Pericarps (μg GAE)	Correlation with Antioxidant Activity *p*-Value
100	468.96 ± 2.95	158.99 ± 11.95	$p < 0.001$
50	248.85 ± 2.59	82.19 ± 6.81	
25	136.56 ± 9.61	31.02 ± 2.09	
10	50.39 ± 1.62	19.70 ± 3.62	
5	30.99 ± 1.32	8.63 ± 0.60	
1	7.22 ± 0.57	1.60 ± 0.55	

[1] GAE, gallic acid equivalent.

All secondary metabolites detected in EMB samples were identified by comparing their chromatographic behaviors and their MS and MS/MS spectra with those of standard reference compounds, when available.

The MS conditions were optimized using reference standards to achieve optimal MS sensitivity for detection and to obtain abundant fragment ions for structural elucidation. Molecules that were identified in negative ion mode belonged to the flavonoid and phenolic acid compound classes. On the other hand, due to the presence of a positive charge in the chemical structure of anthocyanin, good signal sensitivity could also be obtained in positive ion mode.

All compounds were finally confirmed by monitoring their characteristic transitions in MRM mode and comparing their retention times with those of the corresponding authentic standards.

The analytes listed in Table S1 were monitored for their occurrence and 31 compounds were identified in the investigated samples (Table 4).

The precursor/product transitions selected to develop the MRM method are described in Table S1. Quantitative results are reported in Table 4. Each of the three samples was analyzed in triplicate, and the results obtained are expressed as average values of the three analyses.

As shown, ellagic acid was found as the most representative compound in all samples with the highest content in seeds (345 mg/100 g FW), followed by whole EMB (281 mg/100 g FW) and pericarps (244 mg/100 g FW). The other most abundant acids were gallic and quinic acids, ranging 63–123 mg/100 g FW and 77–121 mg/100 g FW, respectively.

With regard to flavonoids, quercetin and quercetin 3-O-rhamnoside were the most abundant (the greatest levels being found in seeds [21 mg/100 g FW and 24 mg/100 g FW, respectively]) followed by isorhamnetin, with values in the range 8–15 mg/100 g FW. Myricetin 3-O-galactoside content was higher in pericarps (10 mg/100 g FW) than in seeds or whole EMB. Overall, the seeds contained the highest level of total polyphenols, at 566 mg/100 g FW. No anthocyanin compounds were found in our samples; this is probably because these compounds are exhaustively extracted during the hydroalcoholic infusion of the myrtle berries in liqueur production. In addition, the low stability of these compounds, which are easily degraded by light, high temperature and air, is widely reported in the literature [43].

Table 4. Polyphenolic contents (mg/100 g DW ± standard deviation) and percentages (%) of different part of exhausted berries of *Myrtus communis*.

Compound	tR	Pericarps		Seeds		Whole EMB		*p*-Value
		mg/100 g ± SD	%	mg/100 g ± SD	%	mg/100 g ± SD	%	
gallic acid	3.27	78.49 ± 5.63	15.02	63.44 ± 4.52	11.22	65.50 ± 9.95	12.70	NS
caffeic acid	9.27	0.07 ± 0.01	0.01	0.06 ± 0.01	0.01	0.07 ± 0.01	0.01	NS
p coumaric acid	10.26	0.51 ± 0.02 [a]	0.10	0.17 ± 0.01	0.03	0.27 ± 0.01 [ab]	0.05	**
ellagic acid	11.42	244.67 ± 14.63 [c]	46.83	345.02 ± 5.95 [a]	61.00	281.79 ± 19.16 [b]	54.64	***
ferulic acid	11.38	0.15 ± 0.00 [b]	0.03	0.20 ± 0.02 [a]	0.04	0.18 ± 0.01 [a]	0.04	***
sinapic acid	11.47	0.02 ± 0.01 [c]	0.00	0.04 ± 0.01 [a]	0.01	0.03 ± 0.01 [b]	0.01	***
quinic acid	1.30	120.82 ± 3.65 [a]	23.13	77.11 ± 1.44 [c]	13.63	96.36 ± 0.60 [b]	18.68	***
siringic acid	9.81	7.48 ± 0.81 [a]	1.43	0.51 ± 0.06 [b]	0.09	2.77 ± 0.08 [ab]	0.54	**
chlorogenic acid	8.93	0.09 ± 0.01	0.02	0.08 ± 0.01	0.01	0.08 ± 0.01	0.02	NS
kaempferol	16.11	1.81 ± 0.04 [b]	0.35	2.07 ± 0.04 [a]	0.37	1.80 ± 0.15 [b]	0.35	*
quercetin	15.42	18.76 ± 0.45 [b]	3.59	20.91 ± 1.01 [a]	3.70	19.18 ± 0.31 [b]	3.72	*
isorhamnetin	16.19	7.67 ± 0.51 [c]	1.47	14.75 ± 0.37 [a]	2.61	9.83 ± 0.38 [b]	1.91	***
myricetin	12.64	7.11 ± 0.31 [a]	1.36	5.26 ± 0.30 [b]	0.93	5.87 ± 0.56 [b]	1.14	**
isoquercetin	11.63	2.30 ± 0.08 [a]	0.44	2.21 ± 0.06 [a]	0.39	1.99 ± 0.05 [b]	0.39	**
quercetin 3-*O*-rhamnoside	12.74	17.24 ± 0.52 [c]	3.30	23.78 ± 0.43 [a]	4.20	19.15 ± 0.26 [b]	3.71	***
robinin	10.77	0.03 ± 0.01 [a]	0.01	0.02 ± 0.01 [b]	0.00	0.02 ± 0.01 [b]	0.00	**
rutin	11.43	0.02 ± 0.01 [ab]	0.00	0.01 ± 0.01 [b]	0.00	0.03 ± 0.01 [a]	0.01	*
isorhamnetin 3-*O*-rutinoside	12.41	0.00 ± 0.00 [a]	0.00	0.00 ± 0.00 [c]	0.00	0.00 ± 0.00 [b]	0.00	***
quercetin 3-*O*-galactoside	11.71	0.32 ± 0.01 [a]	0.06	0.13 ± 0.01 [c]	0.02	0.18 ± 0.01 [b]	0.04	***
myricitrin	11.46	4.62 ± 0.10 [c]	0.88	6.80 ± 0.16 [a]	1.20	5.26 ± 0.28 [b]	1.02	***
neohesperidin	13.11	0.02 ± 0.01 [c]	0.00	0.05 ± 0.01 [a]	0.01	0.03 ± 0.01 [b]	0.00	***
myricetin 3-*O*-galactoside	10.60	9.55 ± 0.06 [a]	1.83	2.65 ± 0.08 [c]	0.47	4.91 ± 0.20 [b]	0.95	***
phloridzin	13.19	0.05 ± 0.01 [a]	0.01	0.04 ± 0.01 [b]	0.01	0.04 ± 0.01 [b]	0.01	***
apigenin	16.03	0.01 ± 0.00 [a]	0.00	0.00 ± 0.00 [b]	0.00	0.00 ± 0.00 [b]	0.00	**
luteolin	15.38	0.01 ± 0.00	0.00	0.01 ± 0.00	0.00	0.01 ± 0.00	0.00	NS
epicatechin	8.72	0.06 ± 0.01 [a]	0.01	0.04 ± 0.01 [b]	0.01	0.04 ± 0.01 [b]	0.01	***
catechin	9.61	0.11 ± 0.01 [a]	0.02	0.09 ± 0.01 [b]	0.02	0.08 ± 0.01 [c]	0.02	***
epigallocatechin	7.68	0.05 ± 0.01 [a]	0.01	0.03 ± 0.01 [b]	0.01	0.04 ± 0.01 [b]	0.01	**
epigallocatechin 3-*O*-gallate	9.28	0.40 ± 0.03 [a]	0.08	0.13 ± 0.01 [c]	0.02	0.21 ± 0.01 [b]	0.04	***
procyanidin B1	7.96	0.02 ± 0.01 [a]	0.00	0.00 ± 0.00 [c]	0.00	0.02 ± 0.01 [b]	0.00	***
procyanidin B2	8.82	0.02 ± 0.01 [a]	0.00	0.00 ± 0.00 [b]	0.00	0.02 ± 0.01 [ab]	0.00	**
cyanidin 3,5-di-*O*-glucoside	9.65	ND		ND		ND		
cyanidin 3-*O*-glucoside	10.50	ND		ND		ND		
cyanidin 3-*O*-arabinoside	11.86	ND		ND		ND		
delphinidin 3-*O*-glucoside	10.00	ND		ND		ND		
malvidin 3-*O*-glucoside	12.01	ND		ND		ND		
pelargonidin 3-*O*-glucoside	13.94	ND		ND		ND		
pelargonidin 3-*O*-rutinoside	14.41	ND		ND		ND		

Means in the same row with different superscripts differ ($p < 0.05$). NS $p > 0.05$; * $p < 0.05$; ** $p < 0.01$; *** $p < 0.001$. ND, not determined (below LOD).

Only few studies have assessed and quantified the polyphenolic composition of myrtle berries: three were focused on whole fresh berries [5,44,45]; one on pericarps [46]; and one specifically looked at the various myrtle berry parts [41]. Thus, a real comparison of our data with other published results is difficult. Nevertheless, the majority of secondary metabolites identified in our samples have previously been reported as present in fresh myrtle fruit; with the exception of caffeic acid, p-coumaric acid, ferulic acid, sinapic acid, quinic acid, syringic acid, chlorogenic acid, isorhamnetin, robinin, isorhamnetin 3-O-rutinoside, neohesperidin, phloridzin, apigenin, luteolin and epicatechin, which were not investigated in the cited papers.

The liquor preparation by hydroalcoholic infusion of berries, extract some of the polyphenolic compounds. Consequently, as expected, the detected levels of the main bulk of polar compounds in EMB were lower than those reported in the literature for fresh myrtle fruit, apart from ellagic acid that was more abundant in our samples. Ellagic acid is a naturally occurring phenolic compound found at high concentrations in many berries; in plants, it forms structural components in the plant cell wall and cell membrane in the form of hydrolysable tannins (ellagitannins), where it is esterified with glucose. Several papers have investigated the biological properties of ellagic acid, which include antioxidant, antimicrobial, anti-inflammatory and antimutagenic activities, as reviewed in [47].

3.4. Antioxidant Activity

The free radical-scavenging properties of the exhausted myrtle berry byproduct are presented in Figure 1, where a lower IC_{50} value (μg/mL) implicates higher antioxidant activity. The ability of DPPH radical scavenging was significantly higher in seeds ($p < 0.01$) than in pericarps, with a three-fold higher antioxidant activity at both time points investigated (0 and 30 min). Our results are in line with those reported by Wannes and Marzouk [41] for the separate myrtle fruit parts, where seeds showed the highest antioxidant activity. This result could be explained by considering the higher content of phenolic acids and flavonols in seeds than in pericarps, as the antioxidant activity of fruit is mainly obtained from phenolic compounds [41].

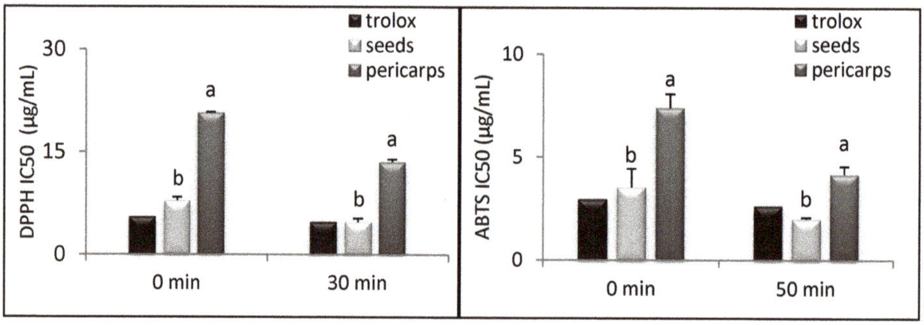

Figure 1. Scavenging of 50% of DPPH and ABTS radical by Trolox and ethanolic extracts from different fruit parts (seeds and pericarps) of exhausted myrtle berries (EMB) at different time points (0 and 30 min). Data were expressed as means ± SD of three independent experiments. Different letters (a,b) indicate significant differences ($p < 0.01$) between seeds and pericarps of EMB at each time point.

The $ABTS\cdot^{+}$ assay showed that antioxidant activity was also significantly higher in seeds ($p < 0.01$) than in pericarps, with values two-fold higher at both time points (Figure 1). A highly significant positive correlation was found by comparing the results obtained using the Folin–Ciocalteau method with the DPPH and ABTS results, respectively (Table 3), confirming the well documented [48] role of phenols in antioxidant activity.

4. Conclusions

Our results demonstrate that exhausted myrtle berries, left over following hydroalcoholic infusion, can still provide a rich source of commercially viable phytochemicals with high antioxidant capacity, carbohydrates, proteins, lipids and polyphenols. These features and the high antioxidant activity of the byproduct support the notion that EMB, in particular the seeds, could be further processed to provide a source of bioactive compounds of bioactive compounds. The possibility of using this byproduct, in its whole form, in feed formulations should also not be excluded.

Supplementary Materials: The following are available online at http://www.mdpi.com/2304-8158/8/7/237/s1, Table S1: LC–MS/MS conditions for quantification of detected compounds by negative/positive ion MRM, Table S2: Accuracy, precision, linearity, LOQ and LOD of LC-ESI-QqQ-MS/MS MRM method for the analysis of standard compounds.

Author Contributions: Formal analysis, writing—original draft preparation, F.C. and M.M.; Formal analysis R.A. and G.L.P.; review and editing, M.P.; data curation supervision, G.B. and G.P. (Giuseppe Pulina); supervision, project administration, A.N.; and review and editing G.P. (Giorgio Pintore).

Funding: This research received no external funding.

Conflicts of Interest: The authors declare no conflict of interest.

References

1. Federici, F.; Fava, F.; Kalogerakis, N.; Mantzavinos, D. Valorisation of agro-industrial by-products, effluents and waste: Concept, opportunities and the case of olive mill wastewaters. *J. Chem. Technol. Biotechnol.* **2009**, *84*, 895–900. [CrossRef]
2. Mauricio, E.M.; Rosado, C.; Duarte, M.P.; Fernando, A.L.; Diaz-Lanza, A.M. Evaluation of Industrial Sour Cherry Liquor Wastes as an Ecofriendly Source of Added Value Chemical Compounds and Energy. *Waste Biomass Valorization* **2018**, 1–10. [CrossRef]
3. Balasundram, N.; Sundram, K.; Samman, S. Phenolic compounds in plants and agri-industrial by-products: Antioxidant activity, occurrence, and potential uses. *Food Chem.* **2006**, *99*, 191–203. [CrossRef]
4. Quideau, S.; Deffieux, D.; Douat-Casassus, C.; Pouysegu, L. Plant polyphenols: Chemical properties, biological activities, and synthesis. *Angew. Chem. Int. Ed.* **2011**, *50*, 586–621. [CrossRef] [PubMed]
5. Tuberoso, C.I.G.; Rosa, A.; Bifulco, E.; Melis, M.P.; Atzeri, A.; Pirisi, F.M.; Dessì, M.A. Chemical composition and antioxidant activities of *Myrtus communis* L. berries extracts. *Food Chem.* **2010**, *123*, 1242–1251. [CrossRef]
6. Quiñones, M.; Miguel, M.; Aleixandre, A. Beneficial effects of polyphenols on cardiovascular disease. *Pharmacol. Res.* **2013**, *68*, 125–131. [CrossRef] [PubMed]
7. Saito, M.; Sakagami, H.; Fujisawa, S. Cytotoxicity and apoptosis induction by butylated hydroxyanisole (BHA) and butylated hydroxytoluene (BHT). *Anticancer Res.* **2002**, *23*, 4693–4701.
8. Anunciato, T.P.; da Rocha Filho, P.A. Carotenoids and polyphenols in nutricosmetics, nutraceuticals, and cosmeceuticals. *J. Cosmet. Dermatol.* **2012**, *11*, 51–54. [CrossRef]
9. Asensi, M.; Ortega, A.; Mena, S.; Feddi, F.; Estrela, J.M. Natural polyphenols in cancer therapy. *Crit. Rev. Clin. Lab. Sci.* **2011**, *48*, 197–216. [CrossRef]
10. Aerts, R.J.; Barry, T.N.; McNabb, W.C. Polyphenols and agriculture: Beneficial effects of proanthocyanidins in forages. *Agric. Ecosyst. Environ.* **1999**, *75*, 1–12. [CrossRef]
11. Grainger, C.; Clarke, T.; Auldist, M.; Beauchemin, K.; McGinn, S.; Waghorn, G.; Eckard, R. Potential use of Acacia mearnsii condensed tannins to reduce methane emissions and nitrogen excretion from grazing dairy cows. *Can. J. Anim. Sci.* **2009**, *89*, 241–251. [CrossRef]
12. Vasta, V.; Yanez-Ruiz, D.R.; Mele, M.; Serra, A.; Luciano, G.; Lanza, M.; Biondi, L.; Priolo, A. Bacterial and protozoal communities and fatty acid profile in the rumen of sheep fed a diet containing added tannins. *Appl. Environ. Microb.* **2010**, *76*, 2549–2555. [CrossRef] [PubMed]
13. Nudda, A.; Correddu, F.; Marzano, A.; Battacone, G.; Nicolussi, P.; Bonelli, P.; Pulina, G. Effects of diets containing grape seed, linseed or both on milk production traits, liver and kidney activities, and immunity of lactating dairy ewes. *J. Dairy Sci.* **2015**, *98*, 1157–1166. [CrossRef] [PubMed]
14. Correddu, F.; Gaspa, G.; Pulina, G.; Nudda, A. Grape seed and linseed, alone and in combination, enhance unsaturated fatty acids in the milk of Sarda dairy sheep. *J. Dairy Sci.* **2016**, *99*, 1725–1735. [CrossRef] [PubMed]
15. Dung, N.T.; Van Binh, D.; Mui, N.T.; Preston, T. Effect of cassava hay supplementation on milk production in lactating goats. *Livest. Res. Rural Dev.* **2010**, *22*, 45.
16. Franco, M.A.; Versini, G.; Saba, R. *Caratterizzazione del Liquore "Mirto di Sardegna Tradizionale"*; Associazione Produttori Liquore Mirto di Sardegna Tradizionale/Confindustria; Litotipografia Valdes: Cagliari, Italy, 1998.
17. Tuberoso, C.I.G.; Barra, A.; Cabras, P. Effect of different technological processes on the chemical composition of myrtle (*Myrtus communis* L.) alcoholic extracts. *Eur. Food Res. Technol.* **2008**, *226*, 801–808. [CrossRef]
18. Van Soest, P.J.; Robertson, J.B.; Lewis, B.A. Methods for dietary fiber, neutral detergent fiber, and nonstarch polysaccharides in relation to animal nutrition. *J. Dairy Sci.* **1991**, *74*, 3583–3597. [CrossRef]
19. AOAC. *Official Methods of Analysis of AOAC International*, 17th ed.; Association of Official Analytical Chemists: Arlington, VA, USA, 2000.
20. AOAC. *Official Methods of Analysis of AOAC International*, 18th ed.; Association of Official Analytical Chemists: Gaithersburg, MD, USA, 2005.

21. Guimarães, R.; Barros, L.; Carvalho, A.M.; Ferreira, I.C.F.R. Studies on chemical constituents and bioactivity of rosa micrantha: An alternative antioxidants source for food, pharmaceutical, or cosmetic applications. *J. Agric. Food Chem.* **2010**, *58*, 6277–6284. [CrossRef]
22. Kramer, J.K.G.; Fellner, V.; Dugan, M.E.R.; Sauer, F.D.; Mossoba, M.M.; Yurawecz, M.P. Evaluating acid and base catalysts in the methylation of milk and rumen fatty acids with special emphasis on conjugated dienes and total trans fatty acids. *Lipids* **1997**, *32*, 1219–1228. [CrossRef]
23. Kramer, J.K.; Cruz-Hernandez, C.; Deng, Z.; Zhou, J.; Jahreis, G.; Dugan, M.E. Analysis of conjugated linoleic acid and trans 18: 1 isomers in synthetic and animal products. *Am. J. Clin. Nutr.* **2004**, *79*, 1137S–1145S. [CrossRef]
24. Brand-Williams, W.; Cuvelier, M.E.; Berset, C. Use of free radical method to evaluate antioxidant activity. *LWT-Food Sci. Technol.* **1995**, *28*, 25–30. [CrossRef]
25. Maldini, M.; Montoro, P.; Addis, R.; Toniolo, C.; Petretto, G.L.; Foddai, M.; Nicoletti, M.; Pintore, G. A new approach to discriminate Rosmarinus officinalis L. plants with antioxidant activity, based on HPTLC fingerprint and targeted phenolic analysis combined with PCA. *Ind. Crops Prod.* **2016**, *94*, 665–672. [CrossRef]
26. Petretto, G.L.; Maldini, M.; Addis, R.; Chessa, M.; Foddai, M.; Rourke, J.P.; Pintore, G. Variability of chemical composition and antioxidant activity of essential oils between *Myrtus communis* var. Leucocarpa DC and var. Melanocarpa DC. *Food Chem.* **2016**, *197*, 124–131. [CrossRef] [PubMed]
27. Lizcano, L.J.; Bakkali, F.; Ruiz-Larrea, M.B.; Ruiz-Sanz, J.I. Antioxidant activity and polyphenol content of aqueous extracts from Colombian Amazonian plants with medicinal use. *Food Chem.* **2010**, *119*, 1566–1570. [CrossRef]
28. Maldini, M.; Foddai, M.; Natella, F.; Addis, R.; Chessa, M.; Petretto, G.L.; Tuberoso, C.I.G.; Pintore, G. Metabolomic study of wild and cultivated caper (Capparis spinosa L.) from different areas of Sardinia and their comparative evaluation. *J. Mass Spectrom.* **2016**, *51*, 716–728. [CrossRef] [PubMed]
29. EMEA; European Medicines Agency. Quality Guidelines: Validation of Analytical Procedures: Text and Methodology (ICHQ2). 1995. Available online: http://www.ema.europa.eu/ema/index.jsp?curl=pages/regulation/general/general_content_000768.jsp&mid=WC0b01ac0580028e8d/ (accessed on 22 September 2016).
30. Free Statistics Calculators. Available online: http://www.danielsoper.com/statcalc3/calc.aspx?id=103 (accessed on 1 February 2017).
31. Nudda, A.; Correddu, F.; Atzori, A.S.; Marzano, A.; Battacone, G.; Nicolussi, P.; Bonelli, P.; Pulina, G. Whole exhausted berries of *Myrtus communis* L. supplied to dairy ewes: Effects on milk production traits and blood metabolites. *Small Rumin. Res.* **2017**, *155*, 33–38. [CrossRef]
32. Nudda, A.; Buffa, G.; Atzori, A.S.; Cappai, M.G.; Caboni, P.; Fais, G.; Pulina, G. Small amounts of agro-industrial byproducts in dairy ewes diets affects milk production traits and hematological parameters. *Anim. Feed Sci. Technol.* **2019**, *251*, 76–85. [CrossRef]
33. Wannes, W.A.; Marzouk, B. Characterization of myrtle seed (*Myrtus communis* var. baetica) as a source of lipids, phenolics, and antioxidant activities. *J. Food Drug Anal.* **2016**, *24*, 316–323. [CrossRef]
34. Wannes, W.A.; Mhamdi, B.; Sriti, J.; Bettaieb, I.; Tounsi, M.S.; Marzouk, B. Fatty acid and glycerolipid changes during Tunisian myrtle (*Myrtus communis* var. italica) fruit ripening. *J. Food Biochem.* **2011**, *35*, 177–194. [CrossRef]
35. Cakir, A. Essential oil and fatty acid composition of the fruits of Hippophae rhamnoides L. (Sea Buckthorn) and *Myrtus communis* L. from Turkey. *Biochem. Syst. Ecol.* **2004**, *32*, 809–816. [CrossRef]
36. Wannes, W.A.; Mhamdi, B.; Marzouk, B. Variations in essential oil and fatty acid composition during *Myrtus communis* var. italica fruit maturation. *Food Chem.* **2009**, *112*, 621–626. [CrossRef]
37. Nguyen, M.T.; Hanzelmann, D.; Härtner, T.; Peschel, A.; Götz, F. Skin-Specific unsaturated fatty acids boost the Staphylococcus aureus innate immune response. *Infect. Immun.* **2016**, *84*, 205–215. [CrossRef] [PubMed]
38. Shingfield, K.J.; Bonnet, M.; Scollan, N.D. Recent developments in altering the fatty acid composition of ruminant-derived foods. *Animal* **2013**, *7*, 132–162. [CrossRef] [PubMed]
39. Ramos, M.J.; Fernández, C.M.; Casas, A.; Rodríguez, L.; Pérez, Á. Influence of fatty acid composition of raw materials on biodiesel properties. *Bioresour. Technol.* **2009**, *100*, 261–268. [CrossRef] [PubMed]
40. Correddu, F.; Nudda, A.; Battacone, G.; Boe, R.; Francesconi, A.H.D.; Pulina, G. Effects of grape seed supplementation, alone or associated with linseed, on ruminal metabolism in Sarda dairy sheep. *Anim. Feed Sci. Technol.* **2015**, *199*, 61–72. [CrossRef]

41. Wannes, W.A.; Marzouk, B. Differences between myrtle fruit parts (*Myrtus communis* var. italica) in phenolics and antioxidant contents. *J. Food Biochem.* **2013**, *37*, 585–594.
42. Correddu, F.; Fancello, F.; Chessa, L.; Atzori, A.S.; Pulina, G.; Nudda, A. Effects of supplementation with exhausted myrtle berries on rumen function of dairy sheep. *Small Rumin. Res.* **2019**, *170*, 51–61. [CrossRef]
43. Montoro, P.; Tuberoso, C.I.; Piacente, S.; Perrone, A.; De Feo, V.; Cabras, P.; Pizza, C. Stability and antioxidant activity of polyphenols in extracts of *Myrtus communis* L. berries used for the preparation of myrtle liqueur. *J. Pharm. Biomed. Anal.* **2006**, *41*, 1614–1619. [CrossRef]
44. Barboni, T.; Cannac, M.; Massi, L.; Perez-Ramirez, Y.; Chiaramonti, N. Variability of polyphenol compounds in *Myrtus communis* L. (Myrtaceae) berries from Corsica. *Molecules* **2010**, *15*, 7849–7860. [CrossRef]
45. Barboni, T.; Venturini, N.; Paolini, J.; Desjobert, J.M.; Chiaramonti, N.; Costa, J. Characterisation of volatiles and polyphenols for quality assessment of alcoholic beverages prepared from Corsican *Myrtus communis* berries. *Food Chem.* **2010**, *122*, 1304–1312. [CrossRef]
46. Martín, T.; Rubio, B.; Villaescusa, L.; Fernández, L.; Díaz, A.M. Polyphenolic compounds from pericarps of *Myrtus communis*. *Pharm. Biol.* **1999**, *37*, 28–31. [CrossRef]
47. Vattem, D.A.; Shetty, K. Biological functionality of ellagic acid: A review. *J. Food Biochem.* **2005**, *29*, 234–266. [CrossRef]
48. Petretto, G.L.; Tuberoso, C.I.G.; Vlahopoulou, G.; Atzei, A.; Mannu, A.; Zrira, S.; Pintore, G. Volatiles, color characteristics and other physico-chemical parameters of commercial Moroccan honeys. *Nat. Prod. Res.* **2016**, *30*, 286–292. [CrossRef] [PubMed]

foods

MDPI

Article

Characterization of Polyphenolic Compounds in Cantaloupe Melon By-Products

Filomena Monica Vella [1], Domenico Cautela [2] and Bruna Laratta [1,*]

1 Consiglio Nazionale delle Ricerche (CNR), Istituto di Ricerca sugli Ecosistemi Terrestri (IRET), via P. Castellino, 111-80131 Napoli, Italy; monica.vella@iret.cnr.it

2 Stazione Sperimentale per le Industrie delle Essenze e dei derivati dagli Agrumi (SSEA), Azienda Speciale della Camera di Commercio di Reggio Calabria, via T. Campanella, 12-89125 Reggio Calabria, Italy; dcautela@ssea.it

* Correspondence: bruna.laratta@cnr.it; Tel.: +39-081-6132329

Received: 2 April 2019; Accepted: 1 June 2019; Published: 6 June 2019

Abstract: The Muskmelon (*Cucumis melo* L.), which includes several crops of great economic importance worldwide, belongs to the Cucurbitaceae family, and it is well recognized for culinary and medicinal purposes. The high fruit consumption produces a large quantity of waste materials, such as peels and seeds that are still rich in molecules like polyphenols, carotenoids, and other biologically active components that possess a positive influence on human health and wellness. A sustainable development in agro-food and agro-industry sectors could come through the reutilization and valorization of these wastes, which in turn, could result in reducing their environmental impact. The current study provides a biochemical characterization of cantaloupe by-products, peels and seeds, through evaluating total polyphenols, *ortho*-diphenols, flavonoids, and tannins content. Furthermore, the antioxidant activity was assessed in order to understand potential benefits as natural antioxidants. Overall, the peel extract revealed the highest radical's scavenging and reducing activities, moreover, it showed higher polyphenolic content than seed extract as revealed by both cromatographic and spectrophotometric analyses. The results of the present study indicate that the melon residues are a good source of natural phytochemicals useful for many purposes, such as ingredients for nutraceutic, cosmetic, or pharmaceutical industries, development of functional ingredients and new foods, and production of fertilizers and animal feed.

Keywords: *Cucumis melo*; polyphenols; flavonoids; antioxidants; by-products; waste valorization

1. Introduction

The Cucurbitaceae family covers several species of great economic importance, including muskmelon (*Cucumis melo* L.), which is largely cultivated and consumed in Europe. Muskmelon encircles a wealth of varietal types, such as smooth-skinned varieties like Honeydew, Crenshaw, and Casaba (*C. melo* var. *inodorous*), rough-skinned varieties like Cantaloupe, Persian melon, and Santa Claus or Christmas melon (*C. melo* var. *reticulatus*), and varieties used when they are immature as vegetables like Barattiere, Carosello, and Armenian Cucumber (*C. melo* var. *flexuosus*). The Cantaloupe melon is well recognized by its net-like slightly ribbed, gray-to-green or light brown skin. It is one of the most consumed melons worldwide thanks to its sweetness, juicy taste, pleasing flavor, and nutritional value [1,2]. In 2016, about 1.9 million tons of melon were harvested in the Mediterranean area, with Spain, Italy, and France representing the main European producers, accounting for 35%, 34%, and 13% overall yield, respectively [3]. In Italy, Cantaloupe is the most cultivated variety. Its name is supposed to derive from Italian "Cantalupo in Sabina", which was formerly a papal county seat near Rome [4].

Cantaloupe is an excellent source of vitamin A, vitamin C, and microelements such as potassium and magnesium [1,4,5]. In recent years, it has been shown to possess useful medicinal properties such as analgesic, anti-inflammatory, antioxidant, antiulcer, anticancer, antimicrobial, diuretic, and antidiabetic properties [2,6,7]. Furthermore, it showed a hepato-protective effect, activity against hypothyroidism and immune-modulator action [6].

Ever-increasing demand for healthy food has stimulated the manufacturing sectors to search for new natural sources of nutritional and healthful components to be employed as food additives or supplements, with high nutritional value [8,9]. As a consequence, the European Union has encouraged the exploitation of fruit by-products for their use as a source of nutritionally and therapeutically functional ingredients to utilize for dietary intake, and as active ingredients in pharmaceuticals and cosmetic industries [10,11].

During fresh consumption and industrial processing of melons (juices, compotes, and salads), large quantities of peels and seeds are produced, and are considered waste. The complete utilization of these by-products could minimize the litter volume, so reducing the environmental impact and the economic costs associated to their disposal. Peels and seeds, in fact, are potential sources of phytochemicals, such as polyphenols, carotenoids, flavonoids, and other bioactive compounds with potential health-promoting effects [12,13]. Among them, polyphenol compounds show antioxidant activity, delaying or inhibiting the oxidation of lipids and other molecules, thus playing an important role in defending cells against free radical damage, a very important way of preventing diseases like cancer and cardiovascular disorders [12–15].

Studies related to melon peels and seeds are scarce; data concerning their whole biochemical characterization are insufficient. Recently, Mallek-Ayadi et al. [5] studied only the phenolic composition and functional properties of peels, concluding that this by-product could be considered as a rich source of carbohydrates, proteins, calcium, potassium, and polyphenols. Fundo et al. [1] characterized the edible and the waste parts of cantaloupe melon, only considering the bioactive compounds showing antioxidant activity. These studies suggested that the valorization of cantaloupe by-products should be encouraged because they are important sources of healthy compounds for food, cosmetics, and nutraceutical products.

On this basis, this research aims to examine the peels and seeds from cantaloupe melon, evaluating total polyphenol, *ortho*-diphenol, flavonoid, and tannin contents. At the best of our knowledge, this is the first time that such a comprehensive investigation, by means of spectrophotometric, chromatographic and in vitro assays, is achieved on both extracts from seeds and peels, in order to explore their potential attitude as natural sources of antioxidants.

2. Materials and Methods

2.1. Reagents and Standards

Sodium carbonate, Folin–Ciocalteu reagent, sodium nitrite, sodium molybdate, aluminum chloride 6-hydrate, 2,2-diphenyl-1-picrylhydrazyl (DPPH), sodium acetate, iron(III) chloride 6-hydrate, 2,4,6-tripyridyl-s-triazine (TPTZ), bovin serum albumin (BSA), standards phenolic acids (gallic, caffeic, chlorogenic, syringic, ferulic, and ellagic acids), flavonoids (rutin, quercetin, kaempferol and isorhamnetin) and the HPLC-grade solvents were purchased from Sigma Chemical Co. (St. Louis, MO, USA). Sodium hydroxide, hydrochloric acid and ethanol were obtained from Carlo Erba Reagents (Milan, Italy).

2.2. Extraction of Bioactive Compounds

Three "Prescott" varieties of cantaloupe melons (*Cucumis melo* L. var. *reticulatus*) were purchased in a local market at commercial ripening stage. Peels were manually removed with a knife and seeds were separated too. Peels and seeds were oven-dried at 37 °C, until constant weight. Randomly chosen samples were utilized for analyses. Both samples were milled with a food mixer (Moulinex, Italy) and kept at −20 °C until extractions were performed. For bioactive compounds recovery, 200 mg of fine powder of cantaloupe peels and seeds were extracted with 10 mL of 95% ethanol (ratio 1:50 w/v) for 6 h at 50 °C in a closed vessel using an ultrasonic bath. The extracts were recovered by centrifugation at 13,000× *g* for 10 min at 4 °C, and dried using a rotary evaporator (IKA RV8, IKA-Werke GmbH & Co, Staufen, Germany).

2.3. Total Polyphenols Content

Total polyphenols were spectrophotometrically determined according to the Folin–Ciocalteu method [16]. In brief, 150 µL of extracts were mixed with 750 µL of Folin–Ciocalteu reagent and 600 µL of 7.5% (*w*/*v*) Na_2CO_3. After incubation, the absorbance was read at 765 nm (UV-Vis spectrophotometer, model DMS-200, Varian, Leini, Italy). Total phenolic amount was calculated by a six point calibration curve obtained with different quantities of gallic acid standard solution ranging from 1.5 to 10 µg. (y = 0.0768 x; R^2 = 0.9909) and the results were expressed as mg of gallic acid equivalents (GAE) per g of extract.

2.4. Ortho-Diphenols Content

Ortho-diphenols content was evaluated by Arnow assay [17]. Briefly, 400 µL of extracts were mixed with 400 µL of 0.5 M HCl, 400 µL of 1.45 M $NaNO_2$—0.4 M Na_2MoO_4 and 400 µL of 1 M NaOH. The absorbance was recorded at 500 nm and *ortho*-diphenols were determined by a calibration curve obtained using caffeic acid as standard. The *ortho* diphenolic content was determined by a calibration curve obtained using a caffeic acid standard solution ranging from 5 to 50 µg (y = 0.0152 x; R^2 = 0.9921), and the results were expressed as mg of caffeic acid equivalents (CAE) per g of extract.

2.5. Flavonoid Content

Flavonoid content in the extracts was determined according to the colorimetric method based on the formation of flavonoid-aluminum compounds [18]. In the assay, extracts were mixed with distilled water and $NaNO_2$. After 5 min, $AlCl_3 \times 6H_2O$ were added and the reaction was stopped by adding 1 M NaOH and distilled water. The absorbance was read at 510 nm and (+)-catechin was used to create the standard curve. (+)-Catechin, from 5 to 100 µg was used to create the calibration curve (y = 0.009 x; R^2 = 0.9940) and the results were expressed as mg of catechin equivalents (CE) per g of extract.

2.6. Tannins Content

Total tannins were assessed as reported by Vella et al. [19] incubating extracts with BSA at 30 °C for 1 h. The supernatant, representing the non-tannin fraction, was collected by centrifugation at 13,000× *g* for 10 min at 4 °C and was analyzed using the Folin-Ciocalteu method. Tannins were determined by difference from the amounts of the polyphenols determined before and after BSA precipitation. Tannins were expressed as mg of gallic acid equivalents (GAE) per g of extract.

2.7. In Vitro Antioxidant Activity

The antioxidant activity of cantaloupe peels and seed extracts were evaluated by means of two in vitro biochemical assays: The Ferric Reducing Antioxidant Power (FRAP) and the DPPH (2,2-diphenyl-1-picrylhydrazyl) radical-scavenging activity.

As reported by Benzie and Strain [20], freshly prepared FRAP reagent were added to the extracts. The absorbance was recorded after 4 min at 593 nm. The antioxidant activity of samples was calculated from a calibration curve with L-ascorbic acid ranging from 0.5 to 5 µg (y = 0.1662 x; R^2 = 0.9918) and the results were expressed as mg of ascorbic acid equivalents (AAE) per g of extract.

The free radical scavenging activity (RSA) of the extracts was assessed according to the procedure of Blois [21]. In brief, different concentrations of peels and seeds extracts were mixed with DPPH methanolic solution. The absorbance reduction at 517 nm of the DPPH was determined continuously. The RSA was calculated as a percentage of DPPH discoloration, using the following equation:

$$\% \, RSA = \left[\frac{[A_{DPPH} - A_s]}{A_{DPPH}} \right] * 100, \tag{1}$$

where A_S is the absorbance of the solution when the extract was added and A_{DPPH} is the absorbance of the DPPH solution. The EC_{50} value was obtained from the graph of %RSA against the extract concentrations in mg/mL.

2.8. Cromatographic Analyses

High performance liquid chromatography-photodiode-array-mass spectrometry (HPLC-PDA-ESI-MS/MS) analyses were performed using a Surveyor LC pump, a Surveyor autosampler, coupled with a photodiode array detector (PDA) Surveyor and a LCQ Advantage ion trap mass spectrometer (Thermo Finnigan, Waltham, MA, USA) equipped with Xcalibur 3.1 software (Thermo Fisher Scientific, Waltham, MA, USA).

A volume of 5 µL was employed for the analysis on a Supelco Spherisorb® ODS2 HPLC Column (250 × 4.6 mm), and the column was thermostatically controlled at 35 °C. The elution was conducted, as already reported [22], by employing 0.3% acetic acid solution (solvent A) and acetonitrile (solvent B). A gradient elution was performed as following: the initial solvent was 90% A and 10% B; the gradient elution was changed from 10% to 20% B in a linear mode for 15 min; this composition was maintained at isocratic flow for 10 min; the solvent B reached 50% in 10 min and from 50 to 90% B in 10 min.

Elution was performed at a flow rate of 0.5 mL/min with a splitting system of 2:8 to the MS detector (100 µL/min) and PDA detector (400 µL/min). Analyses were performed with an electrospray ionization (ESI) interface in the negative mode. The optimization of the instrumental parameters for bioactive compounds was performed by continuous infusion (FIA)-ESI MS/MS analyses. Parameters for analysis were set using negative and positive ion modes, with spectra acquired over a mass range from *m/z* 50 to 1100. The ionization conditions were optimized, and the parameters used were as follows: capillary temperature, 210 °C; capillary voltage, −10.0 V; tube lens offset, −50.0 V; sheath gas flow rate, 60.00 arbitrary units; auxiliary gas flow rate, 20.00 arbitrary units; spray voltage, 4.50 kV; and scan range of *m/z* 150–1200. In the MS/MS experiments, normalized collision energy of 35.0% was applied.

PDA data were recorded with 200–600 nm range, and HPLC/UV chromatograms were acquired at three different wavelengths (226, 284 and 369 nm) according to the absorption maxima of analyzed compounds. Figure 3 shows the 284 nm absorption data for all compounds. For the quantitative analysis of phenolic compounds, a calibration curve was obtained by the injection of different concentrations of each standard. Peak identification of phenolic compounds was performed according to their retention time, UV-Vis and mass spectra.

For the quantification of phenolic compounds by HPLC-UV, a calibration curve was obtained through injection of different concentrations of standard mixture, blending aliquots of different stock individual standards into a 10 mL glass volumetric flask. The standard solutions were prepared in methanol in the range of 0.0025–0.045 mg/mL. The reproducibility of the detector response at each concentration level was evaluated by a triplicate injection of standard mix and expressed as percentage of relative standard deviations (RSD%). The RSDs were expected to be less than 2%. The limits of detection (LOD) were established at a signal to noise ratio (S/N) of 3. The limits of quantification (LOQ) were established at a signal to noise ratio (S/N) of 10. LOD and LOQ were experimentally verified by the nine injections of reference compounds in LOQ concentrations.

2.9. Statistical Analysis

All samples were analyzed in triplicates and the results were expressed as mean ± standard deviation (SD). Means, SD, calibration curves and linear regression analyses (R^2) were determined using Microsoft Excel 2013 (Microsoft Corporation, Redmond, WA, USA).

3. Results and Discussion

The phenolic composition and functional activity of melon residues, peels and seeds, were studied in order to explore their beneficial properties in sight of potential industrial applications. In this contribution, a biochemical characterization was obtained through the evaluation of total polyphenols, *ortho*-diphenols, flavonoids, tannins, and antioxidants by means of photometric assays and by HPLC profiling.

In Figures 1 and 2 polyphenol, *ortho*-diphenol, flavonoid, and tannin contents in cantaloupe peels and seeds, respectively, are reported.

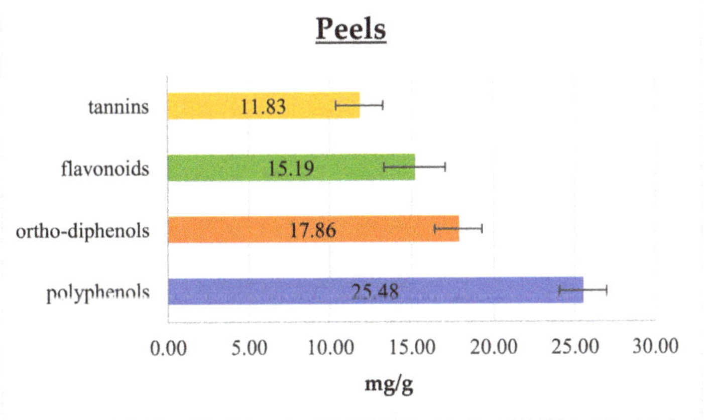

Figure 1. Polyphenol, *ortho*-diphenol, flavonoid, and tannin contents in cantaloupe peels.

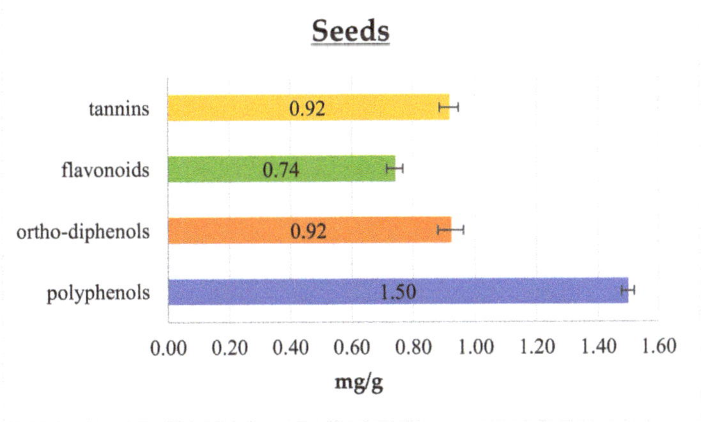

Figure 2. Polyphenol, *ortho*-diphenol, flavonoid, and tannin contents in cantaloupe seeds.

In peels, the polyphenol content was 25.48 ± 1.44 mg GAE/g, which is 6- and 8-fold higher than that reported by Isamil et al. [2] and Mallek-Ayadi et al. [5], respectively. This gap could be attributed to several factors, including cultivar, degree of ripening, and environmental elements, such as climatic conditions and geographical origin [4,19,22]. Conversely, cantaloupe seeds content of polyphenols was 1.50 ± 0.02 mg GAE/g, and this result matches with the range reported in literature data [1,2]. Polyphenols are commonly found in both edible and non-edible plant parts, being essential compounds for their growth and reproduction pathways. They also play an important role in modulating the defense response against insects, pathogens, and microorganisms. Moreover, they take an active part in determining color, flavor, taste, and appearance of fruits. Among phenolics, *ortho*-diphenols are recognized as the most important in relation to their antioxidant activity, since they are able to improve radical stability by forming an intra-molecular hydrogen bond between the hydrogen and phenoxyl radicals. As reported in Figures 1 and 2, *ortho*-diphenols content was 17.86 ± 1.43 and 0.92 ± 0.04 mg CAE/ g in peels and seeds, respectively. To the best of our knowledge, data on *ortho*-diphenols have never been reported in cantaloupe until now.

Considering flavonoids, they are the most common and widely distributed group of plant phenolics, being very effective antioxidants [23]. Cantaloupe peels showed the highest flavonoid content of 15.19 ± 1.88 mg CE/g, meanwhile seeds had 0.74 ± 0.03 mg CE/g, as shown in Figures 1 and 2, respectively. In addition, these components scored higher content values, as well for the total phenolic compounds, than those reported in the literature [2,5]. Moreover, similarly to polyphenols, flavonoid content has many sources of variation such as genotype, fruit ripening, plant phenotypic state and pedoclimatic conditions [4,19,22].

Tannins have been considered health-promoting components of plants, since possessing anti-carcinogenic and anti-mutagenic potentials, as well as antimicrobial, antioxidant and antiradical properties [24–27]. In the present study, tannins content was higher in peels than in seeds, displaying 11.83 ± 1.44 and 0.92 ± 0.03 mg GAE/g, respectively. The literature reports many studies comparing polyphenolic and flavonoid contents of different parts of cantaloupe [1,2,5] but tannins were never assessed.

As the phenolic compounds are known to protect cellular components against free radicals, the antioxidant properties of cantaloupe peels and seed extracts were estimated by means of the FRAP assay and by the DPPH radical scavenging activity, whose results are shown in Table 1. As reported by Benzie and Strain [20], the FRAP assay measures the reduction of a ferric 2,4,6-tripyridyl-s-triazine complex (Fe^{3+}-TPTZ) to the ferrous form (Fe^{2+}-TPTZ) in the presence of an antioxidant compound. The antioxidant ability, measured by the FRAP assay, indicated a higher value in peels as compared to seeds, with values of 12.27 ± 1.22 mg AAE/g and 0.31 ± 0.02 mg AAE/g, respectively.

Table 1. Antioxidants content and activity in cantaloupe peels and seeds.

	Antioxidant Power (mg AAE/g) *	EC_{50} (mg/mL) **
Peels	12.27 ± 1.22	6.65
Seed	0.31 ± 0.02	55.03

* The antioxidant power was measured by FRAP assay; ** EC_{50} was calculated by DPPH assay.

The scavenging activity was studied by means of the DPPH assay, based on the evaluation of the reduction of the DPPH radical to hydrazine as a consequence of the antiradical activity of the extracts. Similarly to the results of the FRAP assay, scavenging activity in the peels extracts was stronger since the EC_{50} after 15 min was 6.65 mg/mL, while seed extracts showed a value of 55.03 mg/mL, thus indicating a lower antioxidant activity of this latter cantaloupe by-product. These results were in agreeance with literature data: in particular, the cantaloupe peels extract in our experiment proved to be 1.4 fold more active when compared to the results published by Isamil et al. [2]. Conversely, seed extract showed an activity that was half than that observed and reported by Isamil et al. [2]. The results of the DPPH assay suggest that extracts are capable of scavenging free radicals via electron or hydrogen-donating mechanisms. Moreover, DPPH activity of these cantaloupe by-products showed similar behavior with the polyphenols, *ortho*-diphenols, flavonoids, and tannins content, thus indicating that radical scavenging activity of cantaloupe peels and seeds extracts is related to the amount of phenolic compounds.

Nowadays, synthetic antioxidants are widely used as additives in food, pharmaceuticals, and cosmetics, but their uses have been questioned because of their possible toxic or carcinogenic activities due to some components formed during their degradation occurring in industrial processing [28].

Therefore, the application of natural plant-based substances may be a suitable alternatives to replace artificial molecules, not only because of their safety, but also since they protect food, feed, and derivatives from the deleterious effects of natural oxidation.

The two extracts from melon peels and seed were also analyzed by HPLC to evaluate the real composition of the phenolic profiles (Figure 3).

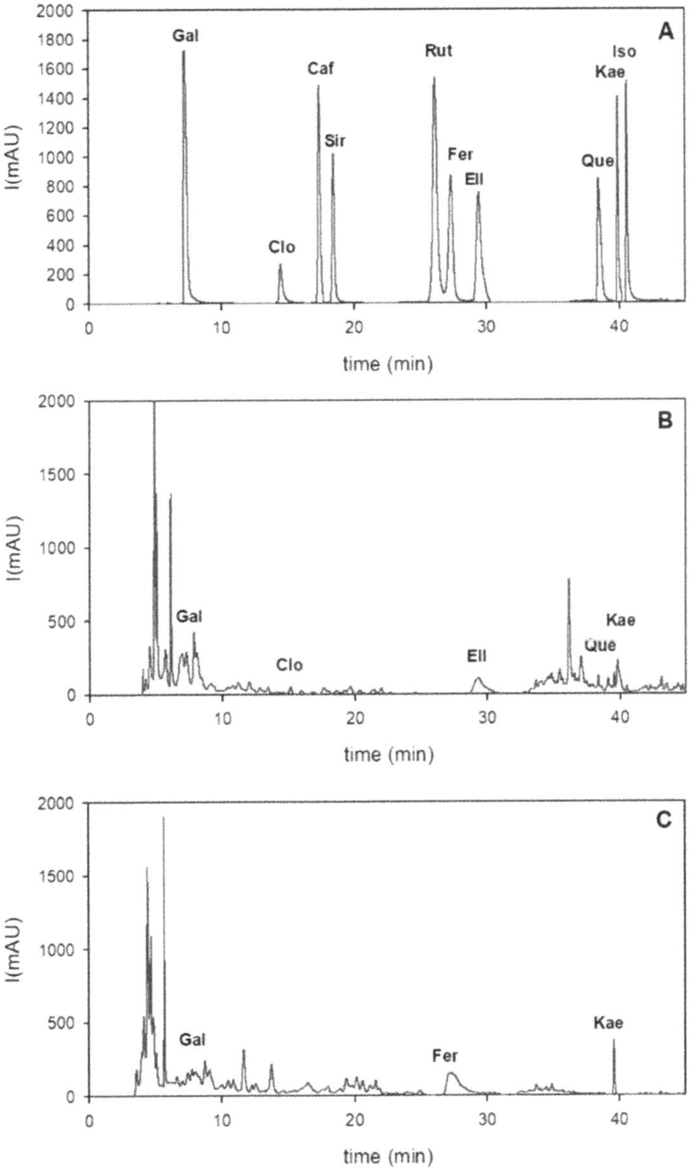

Figure 3. HPLC chromatograms of melon peel and seed extracts, monitored at 284 nm. (**A**) Standard mixture of bioactive compounds. (**B**) Melon peels ethanolic extract. (**C**) Melon seeds ethanolic extract. Gallic acid (Gal), Chlorogenic acid (Clo), Caffeic acid (Caf), Syringic acid (Sir), Rutin (Rut), Ferulic acid (Fer), Ellagic acid (Ell), Quercetin (Que), Kaempferol (Kae), Isorhamnetin (Iso).

Phenolic acids and flavonoids were quantified according to HPLC-PDA data recorded at 284 nm, where gallic acid (2.45 ± 0.08 mg/g), ellagic acid (0.57 ± 0.01 mg/g), and kaempferol (0.32 ± 0.03 mg/g) were the main bioactive compounds found in the peels extract (Table 2; Figure 3B). In seed extract, the richest phytochemicals were ferulic acid (1.51 ± 0.02 mg/g), followed by kaempferol (0.54 ± 0.02 mg/g), and gallic acid (0.07 ± 0.02 mg/g), as reported in Table 2 and in Figure 3C. Peak identification of phenolic compounds was performed according to their retention time, UV-Vis and mass spectra (Table 2).

Table 2. Bio-active contents in melon peel and seed extracts (UV-Vis, HPLC and MS data).

Bioactive Compound	Retention Time (min)	Parent Ion [M-H]-(*m*/*z*)	Transitions (*m*/*z*)	Peel Extract (mg/g)	Seed Extract (mg/g)	$\lambda_{(max)}$
Gallic acid (Gal)	7.39	169.0	125.1	2.45 ± 0.08	0.07 ± 0.02	272
Chlorogenic acid (Clo)	14.62	353.0	191.1	0.08 ± 0.03	0.03 ± 0.01	243,328
Caffeic acid (Caf)	17.52	179.0	135.1	<LOD	<LOD	287,324
Syringic acid (Sir)	18.33	198.1	153.1	<LOD	<LOD	276
Rutin (Rut)	26.15	609.0	299.9	0.06 ± 0.01	nd	258,356
Ferulic acid (Fer)	27.58	193.0	176.9	0.09 ± 0.01	1.51 ± 0.02	287,312
Ellagic acid (Ell)	29.86	301.3	229.0	0.57 ± 0.01	nd	254,364
Quercetin (Que)	38.61	301.0	150.9	0.02 ± 0.01	nd	255,372
Kaempferol (Kae)	39.96	285.0	185.0	0.32 ± 0.03	0.54 ± 0.02	266,366
Isorhamnetin (Iso)	40.51	315.0	286.0	<LOD	<LOD	268,342

LOD = limit of detection S/N: 3 (n = 9) LOD = 0.02 µg/mL.

Gallic acid, ellagic acid and kaempferol have been reported to have antiviral, anti-mutagenic, anticancer, antioxidant and cytotoxic effects [29–31]. Ferulic acid is a ubiquitous natural phenolic compound in seeds; it exhibits a wide variety of biological activities such as antioxidant, anti-inflammatory, antimicrobial, antiallergic, hepatoprotective, anticarcinogenic, antithrombotic, antiviral and vasodilatory actions, increase sperm viability, metal chelation, modulation of enzyme activity, activation of transcriptional factor [31]. Peel and seed HPLC profiles showed some differences from what was reported by Mallek-Ayadi et al. [5] and Zeb [32]. We suppose that these differences might be due to variation relating to different cultivar, environmental conditions during plant growth and fruiting, plant phenotypic state, and possibly due to extraction conditions too [4,19,22].

Altogether, HPLC analysis confirms a higher level of phenolic acids and flavonoids in melon peel than in seed extract. The diverse tissue distribution of bioactive compounds may be attributed to different metabolic roles and networks active in peels and seeds due to their different roles in plant architecture and physiology. Furthermore, it is important to underline that in this study ethanol was used for chemical extraction of peels and seeds and, since it is a GRAS (Generally Recognized As Safe) solvent, it can be utilized safely for bioactive compound recovery, to be used in the food industry.

4. Conclusions

Normally, the non-edible parts of the melon (seeds and peels) are discarded during production processes, reaching approximately 8 to 20 million tons of waste per year worldwide [33]. The extracts of melon exhibit valuable functional and nutraceutical properties, in the light of all the data, spectrophotometry, HPLC profiling and biological activity, obtained in the course of the present study.

With the aim of developing new nutraceuticals, such as supplements, dietary and nutritional products, these cantaloupe by-products seem to be very promising, opening up new perspectives for their use, mainly due to the solubility in water and the stability of their extracts. Therefore, the melon extracts could be used in the production of functional waters, greatly demanded by markets and consumers all over the world, or in food and cosmetic products. Indeed, it has been observed that these by-products act against the oxidation process, thus suggesting their possible future uses as natural colorants and antioxidants in yogurt, biscuits, cupcakes, jellies, sweets and bread, and in anti-wrinkle creams, soaps and bathroom foams, as reported in the literature [33]. Moreover, in this study the reduction to a minimum of the waste volumes and the possibility of developing new products, with the recovery of biomolecules with high added value, may contribute to the sustainable management of waste biomasses that otherwise imply environmental and economic costs.

Author Contributions: Conceptualization, B.L.; formal analysis, F.M.V., D.C. and B.L.; investigation, F.M.V. and D.C.; writing—original draft, F.M.V. and B.L.

Funding: This research received no external funding.

Conflicts of Interest: The authors declare no conflict of interest.

References

1. Fundo, J.F.; Miller, F.A.; Garcia, E.; Santos, J.R.; Silva, C.L.; Brandão, T.R. Physicochemical characteristics, bioactive compounds and antioxidant activity in juice, pulp, peel and seeds of Cantaloupe melon. *J. Food Meas. Charact.* **2018**, *12*, 292–300. [CrossRef]
2. Ismail, H.I.; Chan, K.W.; Mariod, A.A.; Ismail, M. Phenolic content and antioxidant activity of cantaloupe (*Cucumis melo*) methanolic extracts. *Food Chem.* **2010**, *119*, 643–647. [CrossRef]
3. Faostat, Food and Agriculture Organization of the United States. Available online: http://www.fao.org/home/en/ (accessed on 15 February 2019).
4. Maietti, A.; Tedeschi, P.; Stagno, C.; Bordiga, M.; Travaglia, F.; Locatelli, M.; Arlorio, M.; Brandolini, V. Analytical traceability of melon (*Cucumis melo* var *reticulatus*): Proximate composition, bioactive compounds, and antioxidant capacity in relation to cultivar, plant physiology state, and seasonal variability. *J. Food Sci.* **2012**, *77*, C646–C652. [CrossRef] [PubMed]
5. Mallek-Ayadi, S.; Bahloul, N.; Kechaou, N. Characterization, phenolic compounds and functional properties of *Cucumis melo* L. peels. *Food Chem.* **2017**, *221*, 1691–1697. [CrossRef] [PubMed]
6. Milind, P.; Kulwant, S. Muskmelon is eat-must melon. *Int. Res. J. Pharm.* **2011**, *2*, 52–57.
7. Vouldoukis, I.; Lacan, D.; Kamate, C.; Coste, P.; Calenda, A.; Mazier, D.; Conti, M.; Dugas, B. Antioxidant and anti-inflammatory properties of a *Cucumis melo* LC. extract rich in superoxide dismutase activity. *J. Ethnopharmacol.* **2004**, *94*, 67–75. [CrossRef] [PubMed]
8. Galanakis, C.M. Recovery of high added-value components from food wastes: Conventional, emerging technologies and commercialized applications. *Trends Food Sci. Technol.* **2012**, *26*, 68–87. [CrossRef]
9. Schieber, A.; Stintzing, F.C.; Carle, R. By-products of plant food processing as a source of functional compounds—Recent developments. *Trends Food Sci. Technol.* **2001**, *12*, 401–413. [CrossRef]
10. Kosseva, M.R. Processing of food wastes. *Adv. Food Nutr. Res.* **2009**, *58*, 57–136.
11. Laufenberg, G.; Kunz, B.; Nystroem, M. Transformation of vegetable waste into value added products: (A) the upgrading concept; (B) practical implementations. *Bioresour. Technol.* **2003**, *87*, 167–198. [CrossRef]
12. Dai, J.; Mumper, R.J. Plant phenolics: Extraction, analysis and their antioxidant and anticancer properties. *Molecules* **2010**, *15*, 7313–7352. [CrossRef] [PubMed]
13. Moon, J.K.; Shibamoto, T. Antioxidant assays for plant and food components. *J. Agric. Food Chem.* **2009**, *57*, 1655–1666. [CrossRef] [PubMed]
14. Shahidi, F.; Naczk, M. *Phenolics in Food and Nutraceuticals*; CRC Press: Boca Raton, FL, USA, 2004.
15. Sroka, Z.; Cisowski, W. Hydrogen peroxide scavenging, antioxidant and anti-radical activity of some phenolic acids. *Food Chem. Toxicol.* **2003**, *41*, 753–758. [CrossRef]
16. Singleton, V.L.; Rossi, J.A., Jr. Colorimetry of total phenolics with phosphomolybdic-phosphotungstic acid reagents. *Am. J. Enol. Vitic.* **1965**, *16*, 144–158.
17. Arnow, L.E. Colorimetric determination of the components of 3,4-dihydroxyphenylalaninetyrosine mixtures. *J. Biol. Chem.* **1937**, *118*, 531–537.
18. Zhishen, J.; Mengcheng, T.; Jianming, W. The determination of flavonoids contents in mulberry and their scavenging effects on superoxide radicals. *Food Chem.* **1999**, *64*, 555–559. [CrossRef]
19. Vella, F.M.; Laratta, B.; La Cara, F.; Morana, A. Recovery of bioactive molecules from chestnut (*Castanea sativa* Mill.) by-products through extraction by different solvents. *Nat. Prod. Res.* **2018**, *32*, 1022–1032. [CrossRef]
20. Benzie, I.F.; Strain, J.J. The ferric reducing ability of plasma (FRAP) as a measure of "antioxidant power": The FRAP assay. *Anal. Biochem.* **1996**, *239*, 70–76. [CrossRef]
21. Blois, M.S. Antioxidant determination by the use of a stable free radical. *Nature* **1958**, *181*, 1199–1200. [CrossRef]

22. Cautela, D.; Laratta, B.; Santelli, F.; Trifirò, A.; Servillo, L.; Castaldo, D. Estimating bergamot juice adulteration of lemon juice by high-performance liquid chromatography (HPLC) analysis of flavanone glycosides. *J. Agric. Food Chem.* **2008**, *56*, 5407–5414. [CrossRef]
23. Tadmor, Y.; Burger, J.; Yaakov, I.; Feder, A.; Libhaber, S.E.; Portnoy, V.; Meir, A.; Tzuri, G.; Sa'ar, U.; Rogachev, I.; et al. Genetics of flavonoid, carotenoid, and chlorophyll pigments in melon fruit rinds. *J. Agric. Food Chem.* **2010**, *58*, 10722–10728. [CrossRef] [PubMed]
24. Buzzini, P.; Arapitsas, P.; Goretti, M.; Branda, E.; Turchetti, B.; Pinelli, P.; Ieri, F.; Romani, A. Antimicrobial and antiviral activity of hydrolysable tannins. *Mini-Rev. Med. Chem.* **2008**, *8*, 1179–1187. [CrossRef] [PubMed]
25. Koleckar, V.; Kubikova, K.; Rehakova, Z.; Kuca, K.; Jun, D.; Jahodar, L.; Opletal, L. Condensed and hydrolysable tannins as antioxidants influencing the health. *Mini-Rev. Med. Chem.* **2008**, *8*, 436–447. [CrossRef] [PubMed]
26. Santos-Buelga, C.; Scalbert, A. Proanthocyanidins and tannin-like compounds-nature, occurrence, dietary intake, and effects on nutrition and health. *J. Sci. Food Agric.* **2000**, *80*, 1094–1117. [CrossRef]
27. Rechner, A.R.; Kuhnle, G.; Bremner, P.; Hubbard, G.P.; Moore, K.P.; Rice-Evans, C.A. The metabolic fate of dietary polyphenols in humans. *Free Radic. Biol. Med.* **2002**, *33*, 220–235. [CrossRef]
28. Chen, A.Y.; Chen, Y.C. A review of the dietary flavonoid, kaempferol on human health and cancer chemoprevention. *Food Chem.* **2013**, *138*, 2099–2107. [CrossRef] [PubMed]
29. Manach, C.; Williamson, G.; Morand, C.; Scalbert, A.; Remesy, C. Bioavailability and bioefficacy of polyphenols in humans. I Review of 97 bioavailability studies. *Am. J. Clin. Nutr.* **2005**, *81*, 230S–242S. [CrossRef]
30. Tomas-Barberan, F.A.; Clifford, M.N. Dietary hydroxybenzoic acid derivates—Nature, occurrence and dietary burden. *J. Sci. Food Agric.* **2000**, *80*, 1024–1032. [CrossRef]
31. Kumar, N.; Pruthi, V. Potential applications of ferulic acid from natural sources. *Biotechnol. Rep. (Amst.)* **2014**, *4*, 86–93. [CrossRef]
32. Zeb, A. Phenolic profile and antioxidant activity of melon (*Cucumis melo* L.) seeds from Pakistan. *Foods* **2016**, *5*, 67. [CrossRef]
33. Rolim, P.M.; Seabra, L.M.A.J.; de Macedo, G.R. Melon by-products: Biopotential in human health and food processing. *Food Rev. Int.* **2019**, 1–24. [CrossRef]

foods

MDPI

Article

Natural Red Pigment Production by *Monascus Purpureus* in Submerged Fermentation Systems Using a Food Industry Waste: Brewer's Spent Grain

Selim Silbir [1] and Yekta Goksungur [2,*]

1 Engineering Faculty, Department of Food Engineering, Iğdır University, Iğdır 76000, Turkey; selim.silbir@gmail.com

2 Engineering Faculty, Department of Food Engineering, Ege University, Izmir 35040, Turkey

* Correspondence: yekta.goksungur@ege.edu.tr; Tel.: +90-232-311-3027

Received: 24 April 2019; Accepted: 8 May 2019; Published: 11 May 2019

Abstract: This paper studies the production of natural red pigments by *Monascus purpureus* CMU001 in the submerged fermentation system using a brewery waste hydrolysate, brewer's spent grain (BSG). The chemical, structural and elemental characterization of the BSG was performed with Van-Soest method, Fourier-transform infrared spectroscopy (FTIR) and X-ray photoelectron spectroscopy (XPS), respectively. The lignocellulosic structure of BSG was hydrolyzed with a dilute sulfuric acid solution (2% (*w*/*v*)) followed by detoxification with $Ca(OH)_2$. Maximum red pigment production (22.25 UA_{500}) was achieved with the following conditions: 350 rpm shake speed, 50 mL fermentation volume, initial pH of 6.5, inoculation ratio of 2% (*v*/*v*), and monosodium glutamate (MSG) as the most effective nitrogen source. Plackett–Burman design was used to assess the significance of the fermentation medium components, and MSG and $ZnSO_4 \cdot 7H_2O$ were found to be the significant medium variables. This study is the first study showing the compatibility of BSG hydrolysate to red pigment production by *Monascus purpureus* in a submerged fermentation system.

Keywords: Natural red pigment; *Monascus purpureus*; Brewer's spent grain; Submerged fermentation; Plackett-Burman design; Chemical characterization; X-ray photoelectron spectroscopy (XPS); Fourier-transform infrared spectroscopy (FTIR)

---

1. Introduction

In recent years, the food industry has focused on the production of natural pigments from plants and microbial sources to overcome the use of synthetic pigments which are potentially hazardous to human health and the environment. Natural pigments produced by microorganisms have gained more importance because of their low water solubility and the unstable nature of plant-derived pigments against heat and light [1]. Monascus pigments have been used as a natural coloring agent and natural food additive in East Asia. These pigments, which are produced by various species of *Monascus*, improve the color of foods and their sensory characteristics [2,3]. Monascus pigments also have applications as pharmaceuticals since they are reported to have health benefits. Rice fermented with *Monascus purpureus* was found to be effective for the management of cholesterol, diabetes, cardiovascular disease, and also for the prevention of cancer [4].

Monascus is a xerophilic fungus which grows in a wide variety of natural substrates including rice and other cereals [5]. The genus *Monascus* is divided into three species: *purpureus*, *ruber*, and *pilosus*, which are mainly isolated from oriental foods. The most important characteristic of this genus is their ability to synthesize pigments from polyketide chromophores and β-keto acids by esterification. The pigments produced by *Monascus purpureus* are classified into at least six types of pigments based on color: (1) red pigment (rubropunctamin, $C_{21}H_{26}NO_4$, and monascurubramin, $C_{23}H_{27}NO_4$);

(2) orange pigment (rubropunctatin, $C_{21}H_{22}O_5$ and monascorubrin, $C_{23}H_{26}O_5$) and (3) yellow pigment (monascin, $C_{21}H_{26}O_5$ and ankaflavin, $C_{23}H_{30}O_5$). The structure of pigments produced by *Monascus* species depends on factors such as the type of substrate and nitrogen source, pH, temperature, and agitation [2,3,6].

There are various studies on the production of Monascus pigments from food industrial wastes like potato powder [1], bakery waste [2], jackfruit seed [7], grape waste [8], sugarcane bagasse [9,10], wheat [11], sweet potato [12], and prickly pear juice [13].

Brewer's spent grain (BSG) is the most important waste generated by the breweries and corresponds to 85% of the total waste generated in beer production [14]. Since BSG has a low price and is abundant, these properties contribute to the economy of any process that uses this waste biomass. Hence, it will be beneficial to utilize BSG as a substrate in pigment production. There are not many studies on the use of BSG in pigment production in the literature. There is only one paper about pigment production using BSG and it is about solid-state fermentation. Babitha et al. [7], investigated the feasibility of some agro-industrial residues for the production of pigments by *Monascus purpureus*, including brewer's spent grain as a substrate in solid state fermentation. They obtained the best results with jackfruit seed powder and selected this substrate for subsequent studies.

Thus, this is the first study that investigates in detail the use of nutrient-rich brewer's spent grain-derived hydrolysate in the production of red pigments from *Monascus purpureus* CMU-001 in submerged fermentation.

2. Materials and Methods

2.1. Microorganism and Media

Monascus purpureus CMU001 was supplied by Professor Saisamorn Lumyong from Chiang Mai University, Department of Biology. The culture was maintained and sporulated on potato dextrose agar (Merck, Germany). The semi-synthetic fermentation medium was a modification of Silveira et al., [8] and consisted of (g/L): MSG, 8; KH_2PO_4, 5; K_2HPO_4, 5; $MgSO_4{\cdot}7H_2O$, 0.01; $CaCl_2$, 0.01; $ZnSO_4{\cdot}7H_2O$, 0.01; and liquid brewer's spent grain hydrolysate as a carbon source.

Brewer's spent grain (BSG) was supplied by Turk Tuborg Bira ve Malt Sanayi A.Ş. BSG was washed to remove residual starch and its pH neutralized and dried in an oven at 65 °C to 4% moisture content. The substrate was then milled with a pilot scale hammer mill to increase the surface area for better acidic hydrolysis and stored in airtight jars until use.

2.2. Preparation of BSG Hydrolysate

Previously washed, dried and milled brewer's spent grain was treated with different concentrations of sulfuric acid (1–6% (*w*/*v*)) in different solid (BSG): liquid(dilute sulfuric acid) ratios (1:6–10 (w:w)) at 120 °C for 15 min. After hydrolysis, the supernatant was recovered by centrifugation at 6000 rpm for 10 min.

In order to reduce the inhibitory substances generated during acid hydrolysis, a detoxification procedure (overliming) was applied to the raw hydrolysate as recommended by Carvalheiro et al. [15]. The pH of the raw hydrolysate was increased to 10 with $Ca(OH)_2$ and held at 55 °C for 1 h. After centrifugation, the pH of the supernatant was adjusted to pH 5.5 with 25% (*w*/*w*) H_2SO_4 and used as brewer's spent grain-based fermentation medium.

2.3. Cultivation and Fermentation Conditions

The strain was maintained on potato dextrose agar (PDA) dishes, stored at 4 °C, and transferred every 4 weeks to fresh PDA slants incubated at 30 °C for 7 days. Spore solution containing 1.0×10^6 spores/mL was collected using sterile distilled water. Fifty milliliters of semi-synthetic BSG fermentation media was inoculated with 2 % (*v*/*v*) of spore solution. Fermentation experiments were done in a rotary shaker incubator (Sartorius Stedim, Certomat BS-1, Germany) operated at

350 rpm, 30 °C for 7 days. Biomass growth and pigment production were determined at equal time intervals. Different initial fermentation volume (25–75 mL), initial pH (5.5–7.5), inoculation ratio (1–4% (*v*/*v*)), and nitrogen sources (monosodium glutamate, malt sprouts, corn steep liquor, peptone, urea and yeast extract) at a nitrogen concentration equivalent to 8 g/L MSG were tested for Monascus pigment production.

2.4. Analytical Methods

Brewer's spent grain samples were analyzed on the contents of cellulose, hemicellulose, lignin, minerals, and total soluble mass using the Van Soest method [16]. Total solids/Volatile solids (TS/VS) analysis was done to determine the content of total solids, volatile solids and ash according to the standard method 2540 [17].

The infrared spectra of milled brewer's spent grain were recorded between 4000 and 698 cm^{-1} at a resolution of 4 cm^{-1} using Carry 660 FTIR spectrometer (Agilent Technologies, Santa Clara, CA, USA) mounted with a single reflection diamond MIRacle attenuated total reflectance (ATR) accessory a high-pressure clamp (Pike Technologies, Fitchburg, WI, USA). One-hundred-and-twenty-eight scans were co-added for each spectrum to improve the signal-to-noise ratio. The powder was directly placed on a single reflection diamond ATR and pressed with a pressure clamp for having a good contact between the crystal and sample. Four spectra of samples were collected by using a Varian Resolutions Pro 4.05.

X-ray photoelectron spectroscopy (XPS) (Thermo Scientific, K-Alpha XPS, Waltham, MA, USA) was used to determine the elemental composition (C, N, O, P%) of the milled raw material by sending an analytical spot diameter of 200 μm monochromatic Al K-α X-rays (1486.68 eV) to the surface of the sample. Survey and high-resolution spectra of Fe 2p, P 2p, O 1s, N 1s, and C 1s core levels were recorded with a constant pass energy of 200 eV and 50 eV, respectively. A take-off angle of 90° was used in the experiments. In order to detect surface components, XPS measurements were made at different points of the sample after the survey spectra.

The Kjeldahl method [18] was used to determine crude protein content as a function of nitrogen content.

To determine the amount of red pigment extracellularly produced by *M.purpureus*, the mycelia were separated from the fermentation broth using Whatman # 3 filter paper. The filtrate was then centrifuged using a centrifuge (Hettich Universal 320 R, Andreas Hettich GmbH & Co. KG, Tutlingen, Germany). The supernatant was collected to calculate the extracellular pigment concentration by a UV-Vis spectrophotometer (Thermo, Genesys 10S UV-Vis, Waltham, MA, USA) at 500 nm wavelength. Dilution factors were taken into consideration and values obtained were expressed as specific absorbance units (UA_{500}) [8].

Cell growth was evaluated gravimetrically. *Monascus* mycelia were separated and dried to a constant weight at 65 °C and weighed on an analytical scale.

The Dinitrosalicylic(DNS) colorimetric method [19] was used to find the reducing sugar content of the fermentation medium. The reported data are the average values ± standard deviations of three replicates.

2.5. Statistical Analysis

Plackett Burman Design (PB) [20] was applied for screening and understanding the role of various nutrient components in red pigment production using Design Expert Statistical Software (Release 11.1.0.1). MSG, KH_2PO_4, K_2HPO_4, $MgSO_4{\cdot}7H_2O$, $CaCl_2$, and $ZnSO_4{\cdot}7H_2O$ were tested (Table 1) for their effect on red pigment production and the ones with statistically negative or without effect on the red pigment production were excluded from the fermentation medium. The significance of the six examined parameters was evaluated by *p*-value (Table 2). Parameters with a *p*-value smaller than 0.05 were considered statistically significant.

Table 1. Level of variables and experimental design matrix for PB experiment with an observed response.

Code	Variable	Low Level (−)	High Level (+)	#	A	B	C	D	E	F	Pigment Production (UA_{500})
A	MSG	800	8000	1	−	−	+	−	+	+	21.76
B	K_2HPO_4	500	5000	2	+	−	+	+	−	+	17.39
C	KH_2PO_4	500	5000	3	−	+	+	−	+	+	21.25
D	$MgSO_4{\cdot}7H_2O$	10	1	4	+	+	−	−	−	+	7.14
E	$CaCl_2$	10	1	5	+	−	−	−	+	−	12.94
F	$ZnSO_4{\cdot}7H_2O$	10	1	6	−	+	−	+	+	−	5.71
				7	+	+	−	+	+	+	13.94
				8	+	−	+	+	+	−	6.30
				9	−	−	−	−	−	−	15.62
				10	−	−	−	+	−	+	19.32
				11	+	+	+	−	−	−	4.58
				12	−	+	+	+	−	−	20.41

PB: Plackett Burman Design; UA_{500}: absorbance units at 500 nm wavelength.

Table 2. Estimated effects and coefficients (in actual units) for pigment production (UA_{500}).

Term	Effect	Coeff	t	p	% Contribution
Constant		15.146	5.93	0.0000	
A) MSG	−6.96	−9.67	14.38	0.0192	31.57
B) K_2HPO_4	−3.38	−7.51	3.39	0.1395	7.44
C) KH_2PO_4	2.84	6.3×10^{-4}	2.38	0.1976	5.23
D) $MgSO_4{\cdot}7H_2O$	-3.5×10^{-2}	-3.89×10^{-3}	3.63×10^{-4}	0.9857	7.97×10^{-4}
E) $CaCl_2$	−0.43	−0.04744	0.054	0.8276	0.12
F) $ZnSO_4{\cdot}7H_2O$	5.87	0.6526	10.22	0.0330	22.45

$R^2 = 0.9122$.

Plackett-Burman experimental design and the range of variable levels are given in Table 1. There were 12 trials in the experimental design in which low (−) and high (+) levels of each medium component were studied. The following polynomial model explains the relationship between produced red pigment (UA_{500}); Y and fermentation medium components (Equation (1));

$$Y = \alpha_0 + \sum_{i=1}^{n} \alpha_i X_i \tag{1}$$

In Equation (1), Y is the dependent variable (response), α_i is the regression coefficient for linear effects, α_0 is the regression coefficient for the intercept, and X_i is the coded independent variable.

3. Results and Discussion

3.1. The Chemical Composition and Characterization of Brewer's Spent Grain(BSG)

3.1.1. Chemical Composition

The BSG used for pigment production was a lignocellulosic substrate consisting of hemicellulose (53.1%), cellulose (19.2%) and lignin (8.5%), with smaller proportions of ash (3.68%) and nitrogen (2.76%) as shown in Table 3.

It was stated in the literature that the composition of brewer's spent grain is variable. Xiros et al. [21] found that BSG contained 19–40% and 9–25% of hemicellulose and cellulose, respectively. Mussatto and Roberto [22] also found that hemicellulose and cellulose were the main components in BSG followed by lignin. Carvalheiro et al. [23] found that glucan and xylan were the main polysaccharides present in BSG. BSG is a massive waste of barley obtained after the mashing step in the brewing industry. The major constituent of BSG is barley husk which is lignocellulosic in structure. The period of harvest, variety, the mashing and malting conditions employed, and the kinds of supplements added during the brewing process affect the composition of the barley [22].

Table 3. Composition of the milled brewer's spent grain.

Compound	Amount (%)	Analysis/Method
HEMI	53.09	Van Soest [16]
CELL	19.24	
LIGN	8.53	
SOLU	19.15	
Total	100	
Ash	3.68	TS/VS [17]
TS	94.53	
VS	90.05	
Nitrogen	2.76	Kjeldahl [18]
Protein	17.25 *	
Carbon	81.63 ± 1.53	XPS
Nitrogen	2.00 ± 0.59	
Oxygen	15.99 ± 1.23	
Phosphorus	0.66 ± 0.08	

* Protein content determined by multiplying N% with 6.25 conversion factor. HEMI, Hemicellulose; CELL, Cellulose; LIGN, Lignin; SOLU, Soluble solids; TS, Total solids; VS, Volatile solids; XPS, X-ray photoelectron spectroscopy.

3.1.2. Fourier-Transform Infrared Spectroscopy (FTIR)

The infrared spectrum (FTIR) of the milled brewer's spent grain is shown in Figure 1. The broad stretching intense peak at around 3313 cm^{-1} represents hydroxyl and amine groups as shown in the spectrum. The band at 2927 cm^{-1} is related to the asymmetric stretch (nC–H) of–CH2 groups and the corresponding symmetric stretch can be found at 2858 cm^{-1}. The peak at 1743 cm^{-1} can be related to ester bonds or carboxylic (C=O stretching vibration) linkages in lignin and hemicellulose, which is generally evident in untreated BSG and diminishes according to the effectiveness of the pretreatment [24]. Protein-based bonds of nC=O amide I (1635 cm^{-1}) [25] were observed. The peak at 1523 cm^{-1} corresponds to C=C bonds in the aromatic ring of lignin. The peak at 1458 cm^{-1} represents the C–H asymmetric deformation vibrations of aromatic skeletal in lignin [26]. The peak at 1361 cm^{-1} may be associated with highly conjugated C=O stretching vibrations in carboxylic groups. The bands at 1238 cm^{-1} and 1033 cm^{-1} correspond to C–H stretching vibrations, which is characteristic of cellulose content in BSG [24,26]. Parallel results were also found by other researchers [24,26] who used infrared spectroscopy to characterize the chemical structure of BSG by identifying the functional groups present in the samples.

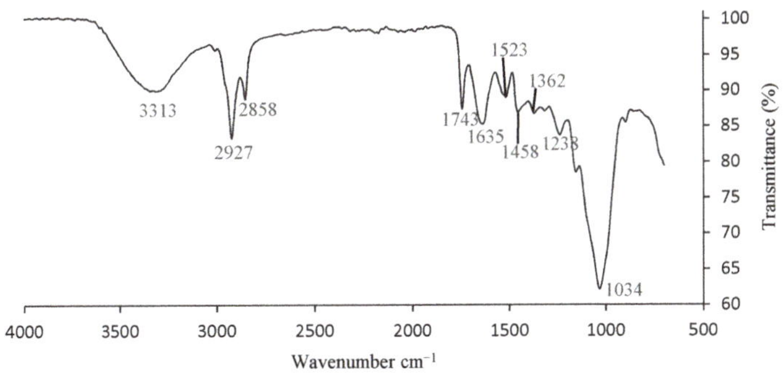

Figure 1. Infrared spectrum(FTIR) of milled brewer's spent grain (BSG).

3.1.3. X-ray Photoelectron Spectroscopy (XPS)

The elemental composition of milled brewer's spent grain samples was examined with XPS. According to the results of XPS analysis, BSG samples contained elements of carbon, oxygen, nitrogen, and phosphorus (Table 4 and Figure 2). The signals from carbon, oxygen, nitrogen, and phosphorus elements are summarized in Table 4. C–C* single bonds (284.99 and 285.3 eV) and C*–OH and C–O–C* single bonds (286.09 eV) indicate the presence of cellulose and hemicellulose in the milled BSG samples. These structures are also supported by signals from the oxygen element of 532.86 eV, 532.4 eV and 532.92 eV which corresponds to the C=O * double bond, C–O*–H single bond and C–O*–H single bond, respectively. The peak of the C=C double bond at 284.36 eV indicates the presence of lignin in brewer's spent grain. The nitrogen signals (C-N*) at 400.11 and 400.44 eV originate from the amine groups in the protein structure of the samples. According to these results and literature data, BSG contains cellulose, hemicellulose, lignin, and protein [26,27].

Table 4. XPS binding energy signals of carbon, oxygen, nitrogen and phosphorus elements.

Element	Binding Energy (eV)	Functional Group
P2p	134.1	P*-O, Phosphate
C1s	285.3	C-C*
N1s	400.11	C-NH_2
O1s	532.92	C-O*-H
C1s	284.99	C-C*
C1s	284.36	C=C*
C1s	286.09	C-OH, C-O-C, C-N
C1s	288.26	C=O
N1s	400.44	C-NH_2
O1s	532.4	C-O*-H
O1s	532.86	C=O*
P2p	133.25	P*-O, Phosphate

XPS: X-ray Photoelectron Spectroscopy. *: excited state.

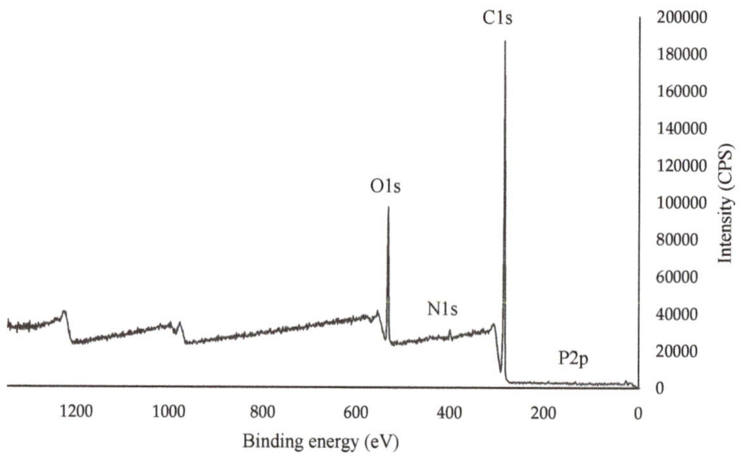

Figure 2. X-ray photoelectron spectroscopy (XPS) survey spectra of milled brewer's spent grain.

3.2. *Effect of Acid Concentration*

The BSG used in this work was a lignocellulosic material composed of hemicellulose (53.1%), cellulose (19.2%) and lignin (8.5%) as shown in Table 3. Hence, a pretreatment step is necessary for the disruption of the lignocellulosic structure of BSG and the release of fermentable monosaccharides like glucose, arabinose, and xylose. Dilute acid hydrolysis is a rapid and basic method employed in the hydrolysis and pretreatment of lignocellulosic substrates such as BSG [15]. The effect of acid concentration on the production of Monascus red pigment was determined using different sulfuric acid concentrations (1–5% (*w*/*v*)). The solid: liquid ratio of 1:8 (w:w), 120 °C and 15 min were used for the pretreatment step. The hydrolysate was then used for pigment production with *Monascus purpureus* CMU001. The fermentation experiments were done in shake flask cultures at 30 °C, pH 5.5 for 7 days. The highest pigment production of 15.35 UA_{500} was obtained in the fermentation medium containing BSG hydrolyzed with 2% (*w*/*v*) sulfuric acid as shown in Figure 3A. When the acid concentration was increased, a gradual decrease in pigment formation was observed. The decrease in pigment formation observed under extreme hydrolysis conditions was probably a result of the production of inhibitory compounds such as HMF and furfural which are known to have toxic effects on microorganisms. In addition, low pigment production (3.07 UA_{500}) was observed in BSG medium hydrolyzed with 1% (*w*/*v*) sulfuric acid. This was probably due to the ineffective disruption of the lignocellulosic structure and inadequate release of fermentable sugars.

This showed that *Monascus purpureus* did not synthesize cellulase to utilize the cellulose moiety of BSG. The rigid lignocellulosic structure of BSG also prevented the utilization of cellulose and therefore pigment production by *Monascus purpureus*.

3.3. Effect of Solid: Liquid Ratio

Three different solid: liquid ratios ((1:6), (1:8) and (1:10) (w:w)) of brewer's spent grain: dilute sulfuric acid were used to hydrolyze BSG using 2% (*w*/*v*) dilute sulfuric acid. BSG hydrolysates obtained were then used for fermentation experiments with *M. purpureus* in shake flask cultures. BSG hydrolyzed with 2% (*w*/*v*) H_2SO_4 using (1:6) (w:w) solid: liquid ratio at 120 °C for 15 min gave the highest value of red pigment production (16.75 UA_{500}) as shown in Figure 3B. Slightly lower pigment production values of 15.35 and 13.10 UA_{500} were obtained with media hydrolyzed at (1:8) and (1:10) solid: liquid ratios, respectively. Solid: liquid ratios lower than (1:6) could not be used due to operational difficulties. These results proved that all the hydrolysates prepared by dilute acid hydrolysis in (1:6), (1:8) and (1:10) (w:w) solid: liquid ratios showed good fermentation characteristics when utilized as a fermentation medium for the red pigment formation by *Monascus purpureus*.

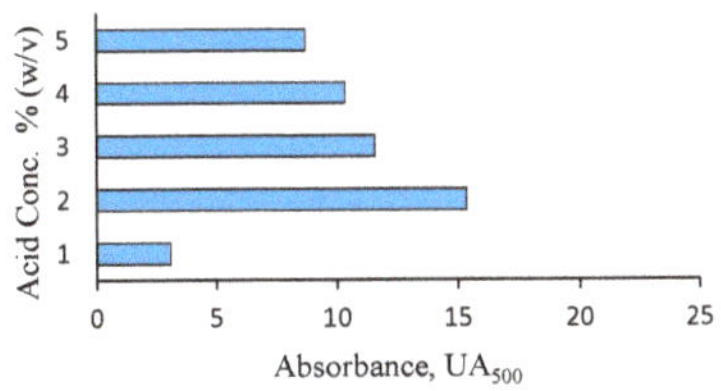

(A)

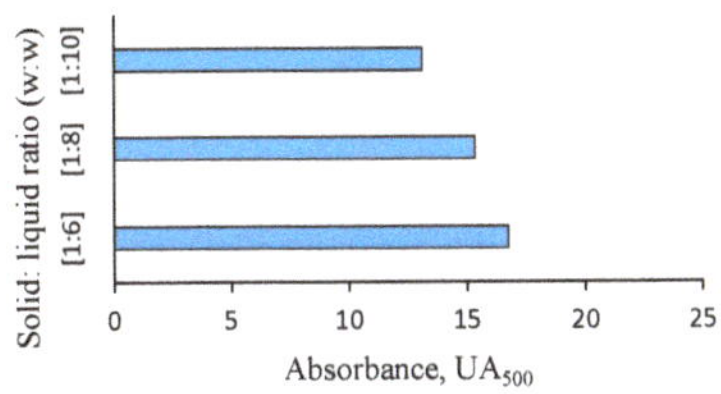

(B)

Figure 3. *Cont.*

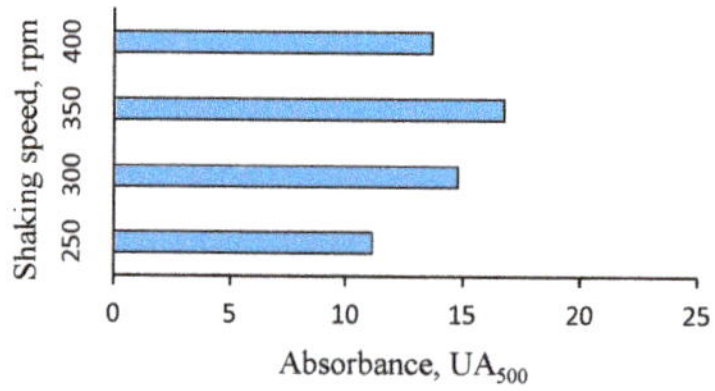

(C)

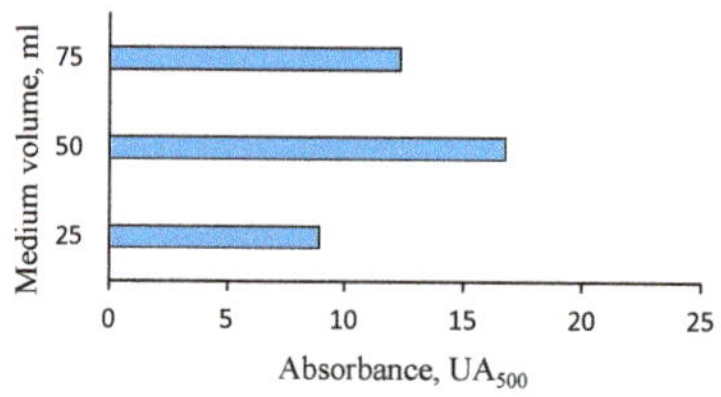

(D)

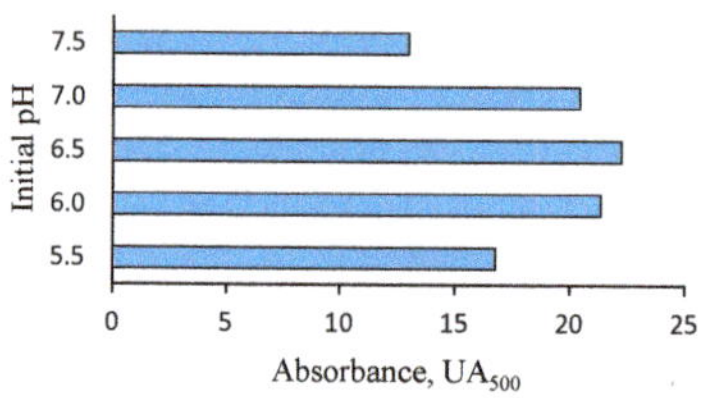

(E)

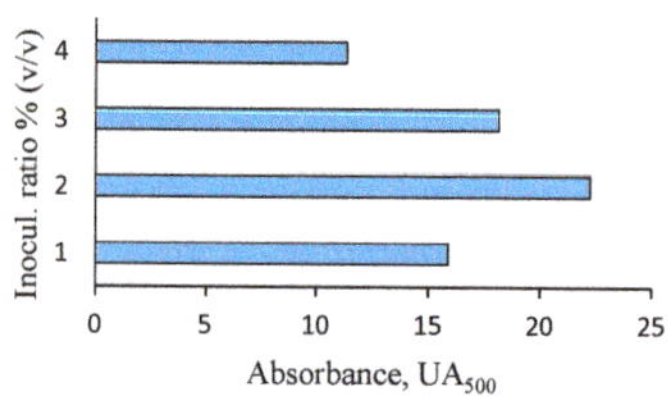

(F)

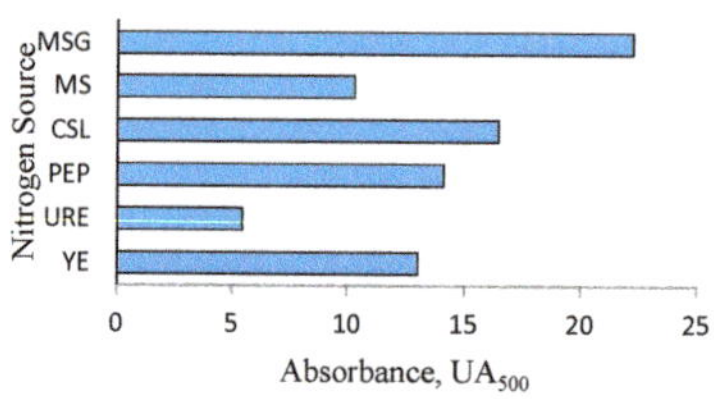

(G)

Figure 3. Effect of various parameters on red pigment production by *Monascus purpureus* CMU 001. (**A**, acid concentration; **B**, solid: liquid ratio; **C**, shaking speed; **D**, medium volume; **E**, initial pH; **F**, inoculation ratio; **G**, nitrogen source (based on equivalent nitrogen amount), (YE, yeast extract; URE, urea; PEP, peptone; CSL, corn steep liquor; MS, malt sprouts; and MSG, monosodium glutamate)). The standard deviation (SD) of each experimental point ranged from ±1.2 to ±4.8.

Musatto and Roberto [28] obtained the highest xylitol production when BSG was hydrolyzed with 100 mg H_2SO_4/g dry substrate. The solid: liquid ratio and the reaction time of sulfuric acid pretreatment was [1:8] (w:w) and 17 min, respectively. Mussatto et al. [14] also used sulfuric acid for BSG hydrolysis prior to lactic acid production by *Lactobacillus delbrueckii*. They employed a sulfuric acid (H_2SO_4) solution of 1.25% in a solid: liquid (g:g) ratio at 120 °C/17 min.

3.4. Effect of Shaking Speed and Medium Volume on Pigment Synthesis

Shaking is crucial in overcoming mass transfer resistances in aerobic fermentation systems and the oxygen transfer rate is directly related with shaker speed [29]. In the present study, the effect of shaking speed and liquid volume on pigment production by *M. purpureus* was studied in BSG medium. Fermentation trials were done in a rotary shaker incubator operated at 250, 300, 350, and 400 rpm. BSG was pretreated with 2% (*w/v*) sulfuric acid solution with a solid: liquid ratio of 1:6 (*w/w*) to prepare the fermentation medium in all experiments. Overliming ($Ca(OH)_2$) was used as the detoxification method. Fermentation lasted for 7 days in 250-mL Erlenmeyer flasks with 50 mL of fermentation medium at pH 5.5. Pigment production was increased with the increased shaking speed (from 250 to 350 rpm) and decreased afterward (Figure 3C). At the lowest shaking speed value of 250 rpm, larger pellets of *Monascus purpureus* were observed, indicating mass diffusion problems. The highest red pigment production of 16.75 UA_{500} was obtained at a shaker speed of 350 rpm. The decrease in red pigment production at a shaker speed of 400 rpm (13.68 UA_{500}) might be due to the shear-sensitive characteristics of *Monascus* mycelia.

To determine the influence of the medium volume in the Erlenmeyer flask on the production of red pigments: 25, 50 and 75 mL of fermentation medium were used in a rotary shaker incubator operated at 350 rpm shaking speed. The highest pigment formation of 16.75 UA_{500} was observed in 50 mL fermentation volume (Figure 3D). The decrease of pigment production in 25 and 75 mL fermentation media volumes might be due to the lower gas liquid mass transfer area resulting in lower oxygen transfer to fermentation liquid and finally to *Monascus* mycelia. Low pigment production of 8.18 UA_{500} in 25 mL medium volume might result from the inefficient shaking of the fermentation medium. The fermentation medium did not rotate well enough with the movement of the orbital shaker which led to lower volumetric mass transfer rates [30]. Vortex formation was noticed in 75 mL of fermentation medium that resulted in the poor mass transfer of oxygen and/or substrate, causing a decrease in red pigment production. Silveira et al. [8] also reported that low oxygen partial pressure decreased *Monascus* pigment production in submerged culture experiments.

3.5. Effect of Initial pH

pH is a significant factor in the activation of important enzymes in pigment production by *Monascus purpureus* [31]. Various initial pH values (5.5, 6.0, 6.5, 7.0 and 7.5) were tested to observe the effect of pH on *M. purpureus* red pigment synthesis. BSG hydrolyzed with 2% (*w/v*) sulfuric acid with a solid: liquid ratio of 1:6 (*w/w*) was used to prepare BSG-based fermentation medium. Fermentation experiments were performed at 30 °C for 7 days. The pH used after sterilization was used as the initial pH since the pH of the medium changed after sterilization. The change in pH after autoclaving might result from the buffering effect of the substrate (brewer's spent grain) or salt solutions.

Initial pH 6.5 gave the highest pigment concentration of 22.25 UA_{500}. Slightly lower red pigment values of 21.38 and 20.43 UA_{500} were obtained at an initial pH of 6.0 and 7.0, respectively (Figure 3E). Lower red pigment production values of 16.75 and 12.98 UA_{500} were obtained at pH values of 5.5 and 7.5, respectively. Our results showed that the synthesis of red pigment by *M. purpureus* slows down with increasing or decreasing pH values at our experimental conditions.

The pH of the medium strongly affects red pigment production in *M. purpureus* since red pigments (extracellular and water soluble) are produced by the chemical modification of orange pigments under relatively higher pH values in the presence of a suitable nitrogen source [32]. Parallel to the findings of our study, many researchers showed that the pH of the fermentation medium had an important effect on pigment synthesis by *M. purpureus*. [31–33].

3.6. Effect of Inoculation Ratio

The biomass and pigment concentration of *Monascus* mycelia are affected by the initial inoculum concentration. To observe the effect of inoculum concentration on pigment production, 50 mL of BSG fermentation medium was inoculated with 1, 2, 3, and 4% (*v*/*v*) spore suspension solution, corresponding to 0.5×10^6, 1.0×10^6, 1.5×10^6, and 2.0×10^6 spores/50 mL of fermentation medium. The highest pigment production (22.25 UA_{500}) was observed in the pretreated BSG medium inoculated with 2% (*v*/*v*) spore suspension. The pigments produced in the fermentation media inoculated with 1, 3 and 4% spore suspensions were 15.87, 18.15 and 11.36 UA_{500}, respectively (Figure 3F). Our results showed that the low inoculum ratio reduced the amount of biomass leading to a lower concentration of pigment. However, the high inoculum ratio yielded a high biomass concentration that resulted in rapid consumption of nutrients in the fermentation medium required for pigment synthesis.

It is well known that spore inoculum concentration affects growth, morphology, volumetric productivity, and enzymes of fungi propagated in submerged culture [34]. However, there are few reports about the effect of inoculum ratio or inoculum size on *Monascus* growth and product formation characteristics. Babitha et al. [7] produced *Monascus* pigments from jackfruit seed by solid-state fermentation and observed poor pigment production at lower and higher inoculum levels similar to the results of this study. They obtained the highest pigment production with an inoculum size of 3 mL (9×10^4 spores/gram dry substrate).

3.7. Effect of Nitrogen Source

The nitrogen source is essential in red pigment production by *Monascus* species since reactions with amino group-containing compounds promote water-soluble, extracellular red pigment production. Monosodium glutamate is a favorable nitrogen source for *M. purpureus* and has been documented by many authors [8,31,32,35,36].

Although MSG is essential in red pigment production, its high cost limits its industrial use. To decrease the cost of pigment production, alternative nitrogen sources were replaced by MSG (8 g/L) on an equivalent nitrogen basis. Fermentation experiments were done in shake flasks inoculated with 2% (*v*/*v*) spore suspension solution at 30 °C, pH 6.5 for 7 days. The batch of each source used was equivalent to a nitrogen amount of 8.0 g/L MSG. The different sources of nitrogen and concentrations used were (g/L): yeast extract (YE), 5.81. peptone (PEP), 4.07; corn steep liquor (CSL), 8.83; malt sprouts (MS), 10.69; and urea (URE), 1.30. Results demonstrated that the nitrogen source greatly influenced the red pigment formation by *M. purpureus*. The highest pigment formation (22.25 UA_{500}) was achieved when MSG was used (Figure 3G). *M. purpureus* produced 16.46 and 13.02 UA_{500} of red pigment when CSL and yeast extract were used as the nitrogen source, respectively. *M. purpureus* also utilized other nitrogen sources; however, lower red pigment production values were obtained. Our results showed that corn steep liquor was a promising alternative source for the production of natural pigments by *M. purpureus*. However, further research is needed to improve red pigment production when CSL is used as the nitrogen source.

3.8. Selecting the Important Medium Components by Plackett-Burman Design

In this research, Plackett–Burman design was employed to test different nutritional variables in pigment production. The aim was to achieve maximum red pigment formation by *M. purpureus*. The six variables tested were MSG, K_2HPO_4, KH_2PO_4, $MgSO_4 \cdot 7H_2O$, $CaCl_2$, and $ZnSO_4 \cdot 7H_2O$ under submerged fermentation. All six factors chosen in the study were tested at two levels, namely low level (−1) and high level (+1), and the experimental runs are given in Table 1. Table 2 shows the estimated regression coefficients, main effect, and *p* and *t* values. Among the chosen variables, MSG and $ZnSO_4 \cdot 7H_2O$ were the significant medium components ($p < 0.05$) (Table 2). Regression coefficients in coded units were used to establish a simple polynomial model in order to estimate pigment production (Equation (2));

$$\text{Red pigment } (UA_{500}) = +13.86 - 3.48*A - 1.69*B + 1.42*C - 0.017*D - 0.21*E + 2.94*F \quad (2)$$

The results obtained in the PB experiment are in agreement with the literature. Bau and Wong [37] investigated the effect of zinc on growth, pigment formation and antibacterial activity of *M. purpureus*. They found that growth, antibacterial activity and pigment production of *M. purpureus* were highly affected by adding zinc to the fermentation medium. They stated that zinc inhibits the growth of *M. purpureus* and increases the production of pigments and antimicrobials. They concluded that zinc might have three roles in *Monascus* metabolism which were promoting glucose uptake, inhibiting mycelial growth and stimulating pigment production.

PB design was also used by other researchers in Monascus pigment production. Sharmila et al. [1] used PB design to evaluate the significant medium variables in Monascus fermentation. They stated that K_2HPO_4, $ZnSO_4{\cdot}7H_2O$ and MSG together with potato powder were significant variables for pigment synthesis by *M. purpureus*. Prajapati et al. [31] used PB design to identify the medium components which affect the pigment formation by *M. purpureus*. They found that tryptone, glucose and pH were significant factors among various variables screened.

3.9. Kinetics of Red Pigment Synthesis by Monascus Purpureus

The kinetics of *M. purpureus* growth together with pigment synthesis were examined in the BSG medium. The fermentation medium was prepared by the dilute sulfuric acid hydrolysis of BSG followed by a detoxification (overliming-unpublished data) step. The medium consisted of dilute sulfuric acid hydrolyzed BSG, MSG (8 g/L) and $ZnSO_4{\cdot}7H_2O$ (0.01 g/L), which were significant medium variables selected from PB screening experiments. Fermentation trials were done in shake flasks at 30 °C, pH 6.5 with a spore inoculation ratio of 2% (*v*/*v*) which corresponds to 2×10^4 spores/mL fermentation medium. The highest pigment production of 22.25 UA_{500} was obtained on the 7th day of fermentation, and afterward it decreased (Figure 4). The oxidative enzymes of the microorganism might be responsible for the oxidation of the pigment, which resulted in a concentration decrease especially during the late period of fermentation. Sharmila et al. [1] obtained the highest pigment production of 6.94 ODU/mL at the 6th day of fermentation by using potato powder as the carbon source. Meinicke et al. [36] obtained 7.38 $UA_{480\ nm}$ of pigment formation after 7 days of fermentation using glycerol as the carbon source. Hamano and Kilikian [35] produced red pigments from a complex culture medium composed of glucose and obtained 20.7 U of pigment production. Haque et al. [2] produced 24 AU/g glucose of pigment using bakery waste hydrolysate. Orozco and Kilikian [33] produced 11.3 U of red pigment from glucose as the carbon source. The variation in the literature data for Monascus pigment production may be related to several factors such as the strain of microorganism, the type of substrate and nitrogen source, the fermentation system, the method of pigment estimation, and the conditions used during fermentation.

On the first day of fermentation, there were no biomass and no pigment formation. First, *Monascus purpureus* mycelia and red color were observed on the second and third day of the fermentation respectively. Maximum pigment production was obtained on day 7 (Figure 5). A maximum biomass concentration of 5.73 g/L was observed after 7 days of fermentation and then it declined. The reducing sugar concentration decreased during fermentation and almost depleted on the 5th day of fermentation. The microorganism probably used non-reducing hydrolysis products of BSG as the carbon source after the 5th day of fermentation.

The pH of the fermentation broth decreased slightly during the first 5 days of fermentation from an initial value of 6.5 to 5.89 and then increased slowly up to 7.74 at the end of fermentation. This was due to the deamination of amino acids present in the medium by *M. purpureus* and the production of ammonia, which increased the pH of the fermentation broth.

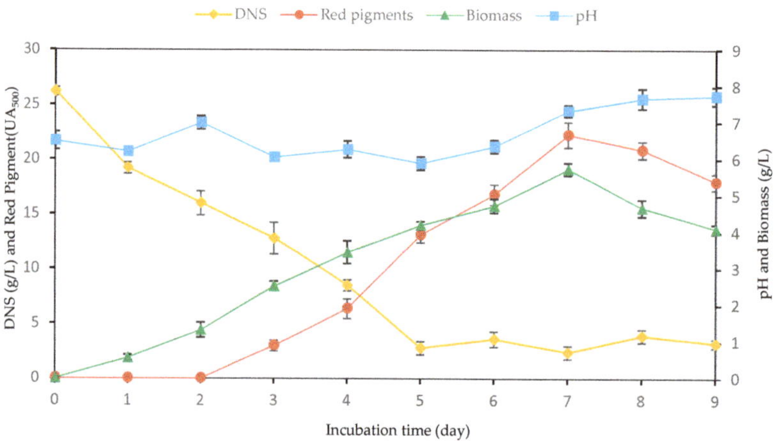

Figure 4. Batch fermentation profile showing red pigment "circles" (UA_{500},), biomass "triangles", pH "squares", and residual sugar "diamonds" for pigment production by *Monascus purpureus* CMU 001. (DNS: dinitrosalicylic colorimetric method; Fermentation conditions: 350 rpm, 30 °C, pH 6.5).

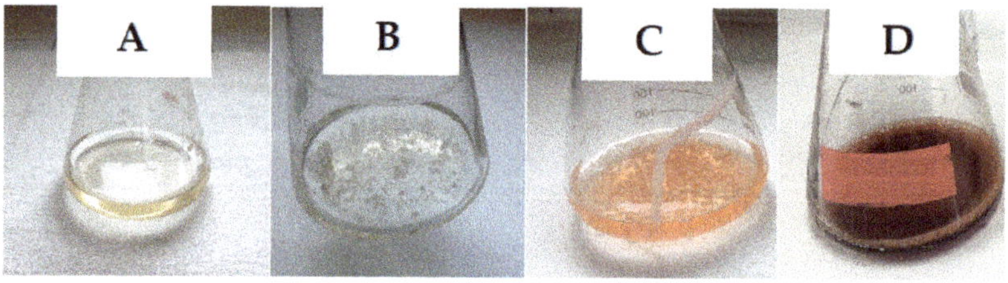

Figure 5. The profile of Monascus mycelia and the produced red pigment in the Erlenmeyer flasks during fermentation (**A**, 1st day; **B**, 2nd day; **C**, 3rd day; **D**, 7th day).

4. Conclusions

Monascus pigments have received worldwide attention for their multiple health benefits; they appear to have anti-mutagenic, anti-cancer, anti-obesity, anti-inflammation, anti-diabetes, and cholesterol-lowering mechanisms. This research article is the first on the evaluation of hydrolyzed and detoxified brewer's spent grain for red pigment production by *Monascus purpureus* in submerged fermentation systems. This paper suggests an innovative approach of using waste produced in large amounts by breweries and contributes to the production of natural pigment products which have multiple benefits for health. Further studies will be focused on the scale-up and stability modeling of the pigments produced.

Author Contributions: Conceptualization, Y.G.; Methodology, S.S. and Y.G.; Validation, Y.G. and S.S.; Data Curation, S.S. and Y.G.; Writing-Original Draft Preparation, S.S.; Writing-Review and Editing, Y.G.

Funding: This research was funded by the EGE UNIVERSITY SCIENTIFIC RESEARCH PROJECT FOUNDATION, 16-MÜH-031.

Acknowledgments: We appreciate Ege University Scientific Research Project Fund (Project No: 16-MUH-031) for supporting this research. The researchers are thankful to Professor Saisamorn Lumyong for providing the *Monascus purpureus* CMU 001, Turk Tuborg Bira ve Malt Sanayi A.Ş for providing brewer's spent grain, N. Altınay Perendeci and Laboratory of Environmental Biotechnology of Akdeniz University for opening their laboratory and supporting the Van Soest analysis and Huriye Goksungur for proofreading the manuscript.

Conflicts of Interest: The authors declare no conflicts of interest. This article does not contain any studies on human or animal subjects. The funders had no role in the design of the study; in the collection, analyses, or interpretation of data; in the writing of the manuscript, or in the decision to publish the results.

References

1. Sharmila, G.; Nidhi, B.; Muthukumaran, C. Sequential statistical optimization of red pigment production by Monascus purpureus (MTCC 369) using potato powder. *Ind. Crops Prod.* **2013**, *44*, 158–164. [CrossRef]
2. Haque, M.A.; Kachrimanidou, V.; Koutinas, A.; Lin, C.S.K. Valorization of bakery waste for biocolorant and enzyme production by Monascus purpureus. *J. Biotechnol.* **2016**, *231*, 55–64. [CrossRef]
3. Kim, H.J.; Kim, J.H.; Oh, H.J.; Shin, C.S. Morphology control of Monascus cells and scale-up of pigment fermentation. *Process Biochem.* **2002**, *38*, 649–655. [CrossRef]
4. Kalaivani, M.; Sabitha, R.; Kalaiselvan, V.; Rajasekaran, A. Health benefits and clinical impact of major nutrient, red yeast rice: A review. *Food Bioprocess Technol.* **2010**, *3*, 333–339. [CrossRef]
5. Babitha, S.; Soccol, C.R.; Pandey, A. Jackfruit seed—A novel substrate for the production of Monascus pigments through solid-state fermentation. *Food Technol. Biotechnol.* **2006**, *44*, 465–471.
6. Dufossé, L.; Galaup, P.; Yaron, A.; Arad, S.M.; Blanc, P.; Murthy, K.N.C.; Ravishankar, G.A. Microorganisms and microalgae as sources of pigments for food use: A scientific oddity or an industrial reality? *Trends Food Sci. Technol.* **2005**, *16*, 389–406. [CrossRef]
7. Babitha, S.; Soccol, C.R.; Pandey, A. Solid-state fermentation for the production of Monascus pigments from jackfruit seed. *Bioresour. Technol.* **2007**, *98*, 1554–1560. [CrossRef] [PubMed]
8. Silveira, S.T.; Daroit, D.J.; Brandelli, A. Pigment production by Monascus purpureus in grape waste using factorial design. *LWT Food Sci. Technol.* **2008**, *41*, 170–174. [CrossRef]
9. Hilares, R.T.; de Souza, R.A.; Marcelino, P.F.; da Silva, S.S.; Dragone, G.; Mussatto, S.I.; Santos, J.C. Sugarcane bagasse hydrolysate as a potential feedstock for red pigment production by Monascus ruber. *Food Chem.* **2018**, *245*, 786–791. [CrossRef] [PubMed]
10. Silveira, S.T.; Daroit, D.J.; Sant'Anna, V.; Brandelli, A. Stability Modeling of Red Pigments Produced by Monascus purpureus in Submerged Cultivations with Sugarcane Bagasse. *Food Bioprocess Technol.* **2013**, *6*, 1007–1014. [CrossRef]
11. Domínguez-Espinosa, R.M.; Webb, C. Submerged fermentation in wheat substrates for production of Monascus pigments. *World J. Microbiol. Biotechnol.* **2003**, *19*, 329–336. [CrossRef]
12. Srivastav, P.; Yadav, V.K.; Govindasamy, S.; Chandrasekaran, M. Red pigment production by Monascus purpureus using sweet potato-based medium in submerged fermentation. *Nutrafoods.* **2015**, *14*, 159–167. [CrossRef]
13. Hamdi, M.; Blanc, P.J.; Goma, G. Effect of aeration conditions on the production of red pigments by Monascus purpureus growth on prickly pear juice. *Process Biochem.* **1996**, *31*, 543–547. [CrossRef]
14. Mussatto, S.I.; Fernandes, M.; Mancilha, I.M.; Roberto, I.C. Effects of medium supplementation and pH control on lactic acid production from brewer's spent grain. *Biochem. Eng. J.* **2008**, *40*, 437–444. [CrossRef]
15. Carvalheiro, F.; Duarte, L.C.; Lopes, S.; Parajó, J.C.; Pereira, H.; Gírio, F.M. Evaluation of the detoxification of brewery's spent grain hydrolysate for xylitol production by Debaryomyces hansenii CCMI 941. *Process Biochem.* **2005**, *40*, 1215–1223. [CrossRef]
16. Van Soest, P.J. Use of Detergents in the Analysis of Fibrous Feeds. II. A Rapid Method for the Determination of Fiber and Lignin. *J. AOAC Int.* **1963**, *46*, 829–835. [CrossRef]
17. Rice, E.W.; Baird, R.B.; Eaton, A.D. (Eds.) *Standard Methods for the Examination of Water and Wastewater*, 23rd ed.; American Public Health Association: Washington, DC, USA; American Water Works Association: Denver, CO, USA; Water Environmental Federation: Alexandria, VA, USA, 2017.
18. Kjeldahl, J.G.C. En ny Methode til Kvaelstofvestemmelse i organiske Stoffer. *Z. Anal. Chem.* **1883**, *22*, 366–382. [CrossRef]
19. Miller, G.L. Use of Dinitrosalicylic Acid Reagent for Determination of Reducing Sugar. *Anal. Chem.* **1959**, *31*, 426–428. [CrossRef]
20. Plackett, R.L.; Burman, J.P. The Design of Optimum Multifactorial Experiments. *Biometrika* **1946**, *33*, 305–325. [CrossRef]
21. Xiros, C.; Topakas, E.; Katapodis, P.; Christakopoulos, P. Hydrolysis and fermentation of brewer's spent grain by Neurospora crassa. *Bioresour. Technol.* **2008**, *99*, 5427–5435. [CrossRef] [PubMed]
22. Mussatto, S.I.; Roberto, I.C. Chemical characterization and liberation of pentose sugars from brewer's spent grain. *J. Chem. Technol. Biotechnol.* **2006**, *81*, 268–274. [CrossRef]

23. Carvalheiro, F.; Esteves, M.P.; Parajó, J.C.; Pereira, H.; Gírio, F.M. Production of oligosaccharides by autohydrolysis of brewery's spent grain. *Bioresour. Technol.* **2004**, *91*, 93–100. [CrossRef]
24. Ravindran, R.; Jaiswal, S.; Abu-Ghannam, N.; Jaiswal, A.K. A comparative analysis of pretreatment strategies on the properties and hydrolysis of brewers' spent grain. *Bioresour. Technol.* **2018**, *248*, 272–279. [CrossRef]
25. Mussatto, S.I.; Dragone, G.; Roberto, I.C. Brewers' spent grain: Generation, characteristics and potential applications. *J. Cereal Sci.* **2006**, *43*, 1–14. [CrossRef]
26. Borel, L.D.M.S.; Lira, T.S.; Ribeiro, J.A.; Ataíde, C.H.; Barrozo, M.A.S. Pyrolysis of brewer's spent grain: Kinetic study and products identification. *Ind. Crops Prod.* **2018**, *121*, 388–395. [CrossRef]
27. Lynch, K.M.; Steffen, E.J.; Arendt, E.K. Brewers' spent grain: A review with an emphasis on food and health. *J. Inst. Brew.* **2016**, *122*, 553–568. [CrossRef]
28. Mussatto, S.I.; Roberto, I.C. Acid hydrolysis and fermentation of brewer's spent grain to produce xylitol. *J. Sci. Food Agric.* **2005**, *85*, 2453–2460. [CrossRef]
29. Liu, Y.S.; Wu, J.Y.; Ho, K.P. Characterization of oxygen transfer conditions and their effects on Phaffia rhodozyma growth and carotenoid production in shake-flask cultures. *Biochem. Eng. J.* **2006**, *27*, 331–335. [CrossRef]
30. Mantzouridou, F.; Roukas, T.; Achatz, B. Effect of oxygen transfer rate on β-carotene production from synthetic medium by Blakeslea trispora in shake flask culture. *Enzyme Microb. Technol.* **2005**, *37*, 687–694. [CrossRef]
31. Prajapati, V.S.; Soni, N.; Trivedi, U.B.; Patel, K.C. An enhancement of red pigment production by submerged culture of Monascus purpureus MTCC 410 employing statistical methodology. *Biocatal. Agric. Biotechnol.* **2014**, *3*, 140–145. [CrossRef]
32. Kang, B.; Zhang, X.; Wu, Z.; Qi, H.; Wang, Z. Effect of pH and nonionic surfactant on profile of intracellular and extracellular Monascus pigments. *Process Biochem.* **2013**, *48*, 759–767. [CrossRef]
33. Orozco, S.F.B.; Kilikian, B.V. Effect of pH on citrinin and red pigments production by Monascus purpureus CCT3802. *World J. Microbiol. Biotechnol.* **2008**, *24*, 263–268. [CrossRef]
34. Tucker, K.G.; Thomas, C.R. Inoculum effects on fungal morphology: Shake flasks vs agitated bioreactors. *Biotechnol. Tech.* **1994**, *8*, 153–156. [CrossRef]
35. Hamano, P.S.; Kilikian, B.V. Production of red pigments by Monascus ruber in culture media containing corn steep liquor. *Braz. J. Chem. Eng.* **2006**, *23*, 443–449. [CrossRef]
36. Meinicke, R.M.; Vendruscolo, F.; Moritz, D.E.; de Oliveira, D.; Schmidell, W.; Samohyl, R.W.; Ninow, J.L. Potential use of glycerol as substrate for the production of red pigments by Monascus ruber in submerged fermentation. *Biocatal. Agric. Biotechnol.* **2012**, *1*, 238–242. [CrossRef]
37. Bau, Y.S.; Wong, H.C. Zinc Effects on Growth, Pigmentation and Antibacterial Activity of Monascus purpureus. *Physiol. Plant.* **1979**, *46*, 63–67. [CrossRef]

MDPI
St. Alban-Anlage 66
4052 Basel
Switzerland
Tel. +41 61 683 77 34
Fax +41 61 302 89 18
www.mdpi.com

Foods Editorial Office
E-mail: foods@mdpi.com
www.mdpi.com/journal/foods